STUDENT UNIT GUIDE

NEW EDITION

Edexcel AS/A2 Biology Units 3 & 6

Practical Biology and Research and Investigative Skills

Ed Lees

Philip Allan, an imprint of Hodder Education, an Hachette UK company, Market Place, Deddington, Oxfordshire OX15 0SE

Orders
Bookpoint Ltd, 130 Milton Park, Abingdon, Oxfordshire OX14 4SB
tel: 01235 827827
fax: 01235 400401
e-mail: education@bookpoint.co.uk
Lines are open 9.00 a.m.–5.00 p.m., Monday to Saturday, with a 24-hour message answering service.
You can also order through the Philip Allan website: www.philipallan.co.uk

ISBN 978-1-4441-8279-8

First printed 2013
Impression number 5 4 3 2
Year 2016 2015 2014

Cover photo: Fotolia

Typeset by Integra Software Services Pvt. Ltd., Pondicherry, India

Printed in Dubai

Contents

Getting the most from this book

Examiner tips

Advice from the examiner on key points in the text to help you learn and recall unit content, avoid pitfalls, and polish your exam technique in order to boost your grade.

Knowledge check

Rapid-fire questions throughout the Content Guidance section to check your understanding.

Knowledge check answers

1 Turn to the back of the book for the Knowledge check answers.

Questions & Answers

Exam-style questions

Examiner comments on the questions
Tips on what you need to do to gain full marks, indicated by the icon ⓔ.

Sample student answers
Practise the questions, then look at the student answers that follow each set of questions.

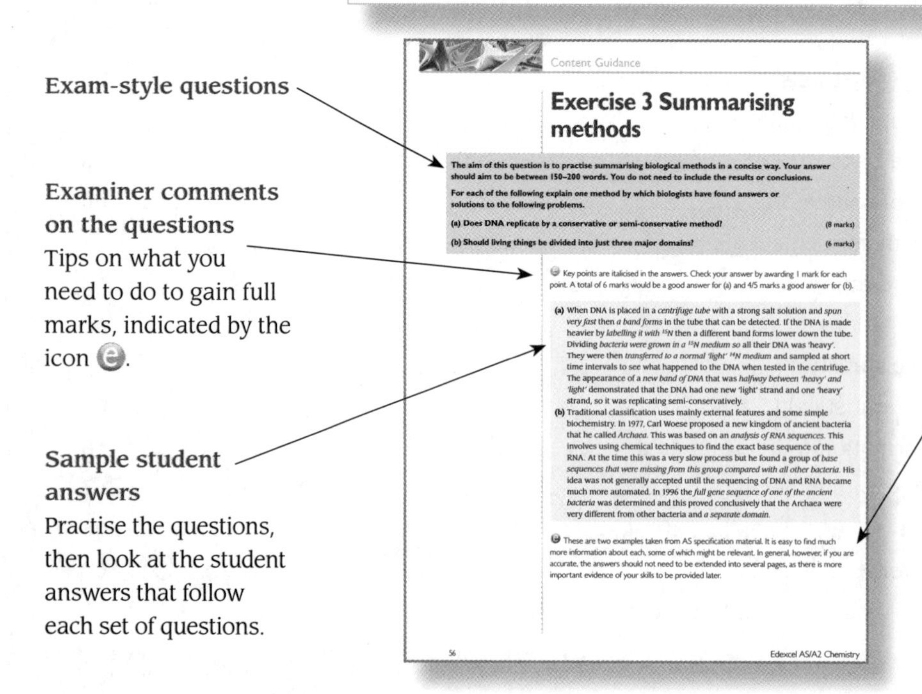

Content Guidance

Exercise 3 Summarising methods

The aim of this question is to practise summarising biological methods in a concise way. Your answer should aim to be between 150–200 words. You do not need to include the results or conclusions.

For each of the following explain one method by which biologists have found answers or solutions to the following problems.

(a) Does DNA replicate by a conservative or semi-conservative method? (8 marks)

(b) Should living things be divided into just three major domains? (6 marks)

ⓔ Key points are italicised in the answers. Check your answer by awarding 1 mark for each point. A total of 6 marks would be a good answer for (a) and 4/5 marks a good answer for (b).

(a) When DNA is placed in a *centrifuge tube* with a strong salt solution and *spun very fast* then *a band forms* in the tube that can be detected. If the DNA is made heavier by *labelling it with ^{15}N* then a different band forms lower down the tube. Dividing *bacteria were grown in a ^{15}N medium so* all their DNA was 'heavy'. They were then *transferred to a normal 'light' ^{14}N medium* and sampled at short time intervals to see what happened to the DNA when tested in the centrifuge. The appearance of a *new band of DNA* that was *halfway between 'heavy' and 'light'* demonstrated that the DNA had one new 'light' strand and one 'heavy' strand, so it was replicating semi-conservatively.

(b) Traditional classification uses mainly external features and some simple biochemistry. In 1977, Carl Woese proposed a new kingdom of ancient bacteria that he called *Archaea*. This was based on an *analysis of RNA sequences*. This involves using chemical techniques to find the exact base sequence of the RNA. At the time this was a very slow process but he found a group of *base sequences that were missing from this group compared with all other bacteria*. His idea was not generally accepted until the sequencing of DNA and RNA became much more automated. In 1996 the *full gene sequence of one of the ancient bacteria* was determined and this proved conclusively that the Archaea were very different from other bacteria and *a separate domain*.

ⓔ These are two examples taken from AS specification material. It is easy to find much more information about each, some of which might be relevant. In general, however, if you are accurate, the answers should not need to be extended into several pages, as there is more important evidence of your skills to be provided later.

56 Edexcel AS/A2 Chemistry

Examiner commentary on sample student answers
Find out how many marks each answer would be awarded in the exam and then read the examiner comments (preceded by the icon ⓔ) following each student answer.

About this book

This guide is to help you prepare for AS Unit 3 Practical Biology and Research Skills (Visit or issue report) and A2 Unit 6 Practical Biology and Investigative Skills (An Individual Investigation). It is divided into three main sections:

- **About this book** — this introduction explains how the different parts of both Unit 3 and Unit 6 are linked and how you can use practical work throughout the course to build your skills.
- **Content Guidance** — this section covers both Unit 3 and Unit 6.
  - Unit 3 — here is an explanation of what is required by each criterion, and also the full background you need before beginning your report. There are supporting questions and exercises to help you understand each point and to develop your skills
  - Unit 6 — here is an explanation of the criteria by which your investigation will be assessed. In each case, important information is summarised to help you progress to A2, and there are exercises and questions to consolidate your progress.
- **Sample Exercises** — this section contains 13 exercises, followed by students' answers and examiner's comments.

How to use this guide

This guide covers both AS Unit 3 and A2 Unit 6, so is relevant to your entire A-level course. Read through the introduction and then select the unit you wish to develop. You will gain most from the exercises and advice in this guide if you use it throughout the whole of your AS or A2 year. Many skills need to be developed over time but you will also gain a great deal by using this guide while completing your coursework.

For each of the criteria by which your report will be assessed there is:

- an explanation of exactly what the criterion requires together with tips on how to achieve high marks
- a summary of factual information or key concepts that you will need to know and understand in order to address the criterion
- exercises and questions to develop your understanding or practise skills

This should allow you to use this book as a basic reference to check that you understand exactly what is required, to find explanations of difficult ideas and to consolidate your learning by answering questions and checking your answers against examiner's comments.

How Science Works

The 'How Science Works' (HSW) section of the specification is made up of 12 criteria and is the most important principle behind the A-level course. You should be familiar with many parts of this as it is also an important part of GCSE courses. The most important feature of these criteria is that they are content-free and are about developing skills and attitudes. If you approach your course as a list of 'facts' to be learned then you are unlikely to do well. Science, and biology in particular, is developing at an increasing pace. Many things thought to be correct even 10–15 years ago are now no longer accepted. So, to progress to AS or A2, it is important not to

Examiner tip

Read the article 'How to read health news' on the NHS website. You can find this at www.nhs.uk/news/Pages/Howtoreadarticlesabouthealthandhealthcare.aspx
It gives an excellent indication of how to develop your scientific thinking and you should read it several times. You will also find an archive of recent articles on this site which are well worth a look.

accept everything in a textbook as a 'fact'. 'How do I know that?' 'Is this the whole story?' are questions you should ask. They form an important part of what is tested in these units.

In addition to skills linked directly to scientific investigations (which we will consider in Unit 6) here are some other things that you will meet throughout your course:

- use of theories, models and ideas to develop and modify scientific explanations
- an appreciation that scientific knowledge is often only the best explanation we have at present and is always being developed and changed
- a consideration of how science is applied to human activities and the benefits and risks this brings
- a consideration of ethical issues in the treatment of humans, other organisms and the environment
- an understanding of how scientists ensure that new knowledge is evaluated critically before being accepted

In the Edexcel specification you will find questions testing HSW skills, not only in Units 3 and 6 but also in the other unit tests. However, **in both Unit 3 and Unit 6 over 90% of the marks are awarded for HSW skills**. Developing these skills is essential if you wish to achieve a high grade.

So how exactly does science work?

The best place to start is to explain how science does not work. The stereotype of white-coated, rather old and strange-looking individuals working away in a laboratory isolated from the real world to reach some 'eureka' moment of a momentous discovery is far from reality. Most research is done by groups of people many of whom are quite young. It is often an international activity.

The starting point for most research is a **scientific model**. This is a theoretical explanation of how and why something happens. To be accepted models need supporting evidence. Scientists gather this evidence by making predictions based on the model and then devising experiments to test these predictions. To do this needs imagination, creativity and a good deal of painstaking work. It might involve using other scientific advances to develop new techniques or using established techniques in an innovative way. As evidence is built up supporting the model it becomes accepted widely and often a more detailed explanation is developed. If evidence from investigation does not support the model then scientists revise the model or develop new ones.

The scientific knowledge built up from this process would be useless if the evidence supporting it was flawed or inaccurate. This is why all research is subjected to detailed critical review before and after publication. Research papers are published in **scientific journals**, each of which concentrates on one area of study. Before a research paper is accepted for publication copies are sent to several experienced scientists working in the same field. These scientists look carefully at the methods used, and the data collected, to check there are no important errors in the methods and that the evidence submitted supports the conclusions made. Only when they are fully satisfied will the paper be published. This is called **peer review**. This is not the end of the process since many scientists around the world will read the paper. It will be discussed and in some cases other laboratories will attempt to repeat the research or develop it further.

Important parts of this process are **scientific conferences**. These are meetings of research scientists working in one specialised field, usually from many different countries. Such conferences allow scientists to discuss recent findings and give details of their current research. Conferences form a vital forum for sharing ideas and developing new ones as well as contributing to the peer review process.

The process by which scientists carry out their investigations follows a clear pattern that will be looked at more closely in the planning section of Unit 6. You should have learned something about this pattern in your GCSE studies; this book concentrates on how to develop this to A2 level.

In theory, this all sounds well organised and logical, but remember that research is carried out by human beings and the history of science is full of stories of personal arguments and rivalries. Reading accounts of Watson and Crick's discovery of the structure of DNA and the intense debates caused by Darwin's *On the Origin of Species* will provide interesting background. The work of Marshall and Warren to identify the importance of the bacterium *Helicobacter pylori* in causing peptic ulcer disease is an interesting story of new ideas overcoming established medical practice. This story is summarised by the timeline available at: en.wikipedia.org/wiki/Timeline_of_peptic_ulcer_disease_and_Helicobacter_pylori and in High, N. (2010) 'Stomach Ulcers', *Biological Sciences Review*, Vol. 23, No. 2, pp. 10–12.

Knowledge check 1

Why did certain groups of people oppose the suggestion that ulcers are caused by a bacterium rather than excess acid?

Knowledge check 2

Is the idea that bacteria are the cause of ulcers the full story?

This is a good illustration of how scientific research does not always follow a smooth or conventional pattern. The timeline also shows the role of conferences and how research quickly becomes an international activity.

Core practicals

Each unit has a number of core practical exercises that you are expected to complete. These are not assessed as part of an AS/A2 award, but you and your teacher have to sign a verification form confirming that you have undertaken all the practicals. The most important part of this form is the list of skills addressed.

List of skills to be addressed in core practicals

(1) (a) Handle apparatus correctly and safely
(b) Work safely with due consideration for the wellbeing of living organisms and the environment

(2) (a) Measure and observe precisely and record in a structured manner, identify variables and justify validity and reliability of results
(b) Identify and explain possible systematic or random errors in results

(3) (a) Use appropriate methods to analyse results, present data and identify trends or patterns
(b) Describe anomalies, evaluate methodology and make suggestions to improve or extend the investigation.

In most cases you will carry out these practical exercises in class. It is likely that you will have a sheet of instructions and some apparatus to use to collect data. However, time is often short and it is not always easy to set things up without practice. As a

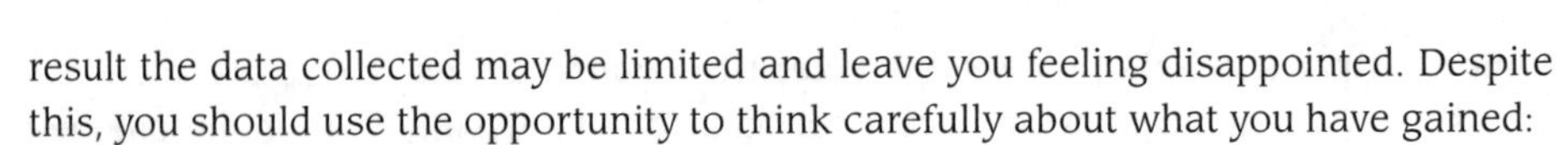

result the data collected may be limited and leave you feeling disappointed. Despite this, you should use the opportunity to think carefully about what you have gained:

- Handling the apparatus and thinking about the variables involved will make it much easier to learn the detail that could be part of a unit test question.
- You will have learned another technique that could be useful when planning your Unit 6 investigation.
- You could concentrate on any one of the skills listed and use the practical to develop your ability.
- You should think carefully about the biological principles behind the technique you have used. Just because they are in the specification does not mean that they are perfect or without major drawbacks.

If you do this repeatedly you will be in a much better position to carry out your own investigation in Unit 6.

Thinking carefully about a core practical

Unit 2 Topic 4: describe how to investigate the antimicrobial properties of plants

A common way of carrying out this practical is to spread a bacterial culture on an agar plate and then place filter paper discs soaked in plant extracts on the plate. During incubation the bacteria grow and cover the plate except for clear zones around the discs where the plant extract has killed the bacteria. The theory suggests that the bigger the clear zone the greater is the antimicrobial effect of the extract.

To develop your skills to A-level standard you should be thinking along these lines:

- What might be the most useful measurement — radius or diameter of the zone? Would area be better?
- If I measure area should I include the area of the disc?
- How is the extract transferred from the disc to the agar?

It is the process that is important here, but for information, area would be the best measurement to take. You should exclude the area of the disc as transfer begins at its edge. The extract moves from the disc by diffusion through an aqueous medium (agar). When comparing different extracts it is important to think about concentration because this changes the rate of diffusion.

Examiner tip

When carrying out any practical exercise think carefully about the science involved. Don't just follow the instructions.

Knowledge check 3

Why is it important to measure initial rate when investigating the rate of an enzyme-catalysed reaction?

Knowledge check 4

Why would drinking coffee or Red Bull not give useful information about the effect of caffeine on heart rate?

Research skills

These skills are an important part of both Unit 3 and Unit 6. There are some exercises and advice in the section on Unit 3 but this is equally important for Unit 6. Many students use the internet as their main source of information but you are required to use other sources too. Above all it is important that, from the first week of your course, you ask questions such as, 'Where is the evidence?' or 'Is this source reliable?' Remember that what may be described in a newspaper headline or on an online social media site might not be acceptable to a group of scientists.

Content Guidance

Unit 3 Practical biology and research skills

Visit or issue report

Basic requirements

- The visit or issue report is not linked to any specific biological content so you can choose a topic that really interests you.
- The report should be an **analytical** piece of work, rather than an essay, and it should not contain long descriptive passages.
- The recommended length is between 1500 and 2000 words. There is no penalty for exceeding this length but writing more does not necessarily mean you will gain more marks.
- The report must be your own original work. Direct quotes must be acknowledged clearly but can only be awarded limited credit. Your report must provide evidence of your individual skills and therefore it must be significantly different from those of your classmates.
- The skills required for this unit need to be developed and practised from the beginning of your course. You will find some suggestions in later sections of this book.

Visit or issue?

This is a question that you must discuss with your teachers. Your school or college may have made arrangements, but you may be able to make your own. However, this will depend on your choice of topic.

Examiner tip
If you choose a visit you must use it to find information and ask questions. This means that you must be well prepared and have a clear idea of what you wish to find out before you go.

Choosing your topic

- Try to choose something that interests you personally.
- Biology-based news items are often a good place to start.
- Refine your title to include a clearly defined problem that biologists are trying to solve.
- Avoid general titles, such as 'global warming' that are too broad.
- Compare your idea with the criteria. Does it seem to offer opportunities to meet them?
- Do some initial research to discover if it is possible to find:
  - the biological background to the problem
  - details of what the biologists are doing to find a solution
  - any evidence of their success
  - details of any alternative solutions

You will have to work hard on these so don't be put off, but if any are impossible then you might need to choose another topic.

Researching information

Details of exactly what is required and how marks are awarded are to be found under criterion 3 later in this book. However, you need to begin your research at the start of your work on Unit 3, particularly to help you check that your title is suitable. Some basic advice on searching for information is included here but you are advised to check some of the details in criteria 3.1 and 3.2 on pp. 19–21.

Examiner tip

Quoting references simply because they contain key words is given no credit. Always check they have a clear link to your hypothesis and give evidence that you are actually using the information.

Knowledge check 5

There is a lot of information to show that manuka honey has antibacterial effects. What are the main problems in translating this into an effective treatment for infections?

Searching web-based sources

Keep it relevant

The internet is an excellent resource but is often used without careful thought. You will be expected to go beyond the first few websites that appear on a Google search. Remember what is placed at the top of a search list is likely to be there not because it is the most useful, but because it has paid the most money.

To avoid having to search through literally millions of irrelevant sites consider using some of the smart search options available on all search engines.

- Do not search for sources using single words or general phrases. Enclose accurate phrases linked to the title of your report in double quote marks.
- If you want to include some alternatives then use 'or' between them to ensure you search for exactly those options.
- Beware of simple question-and-answer sites. You have no idea of the scientific credibility of the person who is posting answers. They could have less knowledge than you!

Finding academic information and data on the web

Gaining access to academic journals with original research information is difficult. Subscriptions are expensive and so such sources are unlikely to be available to you. In addition, most research papers contain technical details that are well beyond A-level — they often require a postgraduate level of understanding. However, abstracts of papers are often available free. An abstract is a short summary of the outline methods and the main findings of the research. This is usually much easier to understand and may well provide you with useful information to quote. The abstract should enable you to decide quickly if it is of any use to your research.

A useful way to begin a web search is to use **Google Scholar**, but remember to enter accurate searches as described in the previous section. This site also contains help in refining searches. Read carefully and you will find all the details you need.

There is an increasing trend for scientific papers to be published free on the internet. The largest of these is the PubMed site, which you can find at www.ncbi.nlm.nih.gov/pubmed. This is a free government sponsored site of academic papers that is well-worth investigating. Although more limited, you will also find references at the end of each section of information on Wikipedia. Clicking on these can lead you to useful information.

Look carefully at the next website that you visit. Can you find out exactly who wrote it? Websites cost a lot of money to set up and maintain and most are not available just for your interest. Does the site have advertising? Does it try to get you to spend money? Will this influence the information it contains?

Non-web-based sources

There are many sources of useful information other than the internet. These might include books, articles, magazines such as *Biological Science Review* and *Scientific American* which have credible academic contributors. You might also be able to obtain useful information from knowledgeable individuals. If you are making a visit then the individuals you meet will be an invaluable resource, so be prepared to use them.

Knowledge check 6

Visit the website at **www.garlic-central.com/antibiotic.html**. Make a list of four features of this web page that might indicate it has low scientific credibility.

Criterion 1: Biologists working to solve a problem

What is required in this criterion?

1.1 Identifying and describing a biological problem or question (4 marks)
1.2 Describing the methods or processes used by biologists to find solutions or data relevant to this question (4 marks)
1.3 Explaining how the methods or processes are appropriate to produce reliable data or solutions using evidence (4 marks)

Getting started

Having selected your title and done some initial research, check carefully that you will be able to meet all the criteria in your report. This is important if you are to avoid wasting time — it would be annoying to find at a later stage that you cannot meet some of the criteria. Do remember that this is meant to give you an opportunity to research any area of biology that interests you; your report should be an interesting read for others too. Biologists are fortunate in having a wide range of topics to choose from. It is rare for a week to go by without an interesting biology-based topic appearing in the national news.

Before going further, make sure that you can answer 'yes' to the following questions:

- Have you identified a clear biological question or problem that will be the main focus of your report?
- Has your research given you confidence that you can explain clearly what biologists are doing to find answers to this question or problem?
- Does your research provide evidence in the form of data, graphs or illustrations of how successful or limited the work of the biologists has been?

If you answer 'no' to any of these questions then you should undertake more detailed research or revise your choice of topic.

If you are using a visit as the basis for your report then these questions are the areas you must be prepared to ask about. The people you meet on your visit are your

most valuable resource. They may be able to suggest sources of information or provide data that you will find helpful — so be prepared, you might not get another chance.

Before you start to write it is also important that you understand what will *not* be helpful if you wish to achieve high marks:

- This is *not* an essay or a 'project' and so should not be a long description of biological information.
- It is *not* an exercise in finding information that you simply copy out.
- You *cannot* gain credit for simply including words and phrases without some explanation or reasoning of your own.

1.1 Identifying the problem or question

What is required in 1.1?

- A clear statement of the precise problem or question your report is about.
- This problem or question must be biological, but not necessarily linked to the specification.
- There must be a description that explains exactly what the problem or question is, and the biology behind it.

This is an important part of the whole report. If you do not have a clear focus then you will find it increasingly difficult to achieve high marks in other criteria. Your focus must be a well-defined question or problem that biologists are working to answer, otherwise you will not be able to address many other criteria.

A simple case study

A student decides to choose 'Gene therapy' as the report title.

Some initial research shows that it is easy to find information, but this is because it is part of the Unit 1 specification material. Such information will only lead to describing the process and will not meet the criteria.

Further research, including an article in *Biological Sciences Review* (April 2012) provided a reference to a US Department of Energy website:

www.ornl.gov/sci/techresources/Human_Genome/medicine/genetherapy.shtml

This site had a lot more information and additional links that allowed the student to develop the title 'Can gene therapy be used to cure inherited blindness?'.

Examiner tip
Avoid choosing titles that are too close to topics in the AS specification. You cannot gain credit for repeating material you have covered during your course.

This case study illustrates how careful research is needed to select a suitable title and how following some useful lines of research can help to provide original ideas.

Examiner tip
Identifying the problem clearly and describing fully do not mean writing a great deal more. Aim to be accurate and concise.

How marks are awarded in 1.1

1.1a	Weak, partial identification of the question or problem	(1 mark)
1.1a	Clear identification of the problem	(2 marks)
1.1b	Question or problem described partially	(1 mark)
1.1b	Question or problem described fully	(2 marks)

1.2 Describing how the problem was solved

What is required in 1.2?

- This section must describe clearly the work of biologists and what they are doing to solve the problem or to answer the question you have identified.
- Make sure you describe the work of biologists, *not* just the technical details of the solutions.
- Research is needed to find some data or evidence of the results of the work you have described that indicate an answer or solution has been found.

When describing methods and processes you need to show that you understand the science involved. A concise explanation is needed, in your own words, not just a description copied from sources. You can use quotes from your research if you acknowledge them, but they must be used as part of your discussion, not as a series of 'cut and paste' items.

While it is often quite easy to find details of how the problem or question is being investigated for 1.2a, it is often more difficult to find evidence for 1.2b to illustrate why the work might be producing useful answers to the problem. If you are basing your report on a visit then this is a key question for the people you meet. Be prepared to ask suitable questions and be ready to take notes on their answers.

This evidence does not necessarily have to be extensive with large amounts of data. It can be in many forms as long as it is relevant to the question or problem.

Examiner tip

Research must go beyond typing a few words into a search engine. Make sure you can find some data or other evidence before you start.

How marks are awarded in 1.2

1.2a	Methods and processes described partially	(1 mark)
1.2a	Methods and processes described fully and explained	(2 marks)
1.2b	A brief description of some of the solution or some limited data	(1 mark)
1.2b	Data or solutions described fully with some explanation	(2 marks)

Note the same pattern here as for the award of marks in 1.1. Simple statements gain only 1 mark; some concise comment or further explanation gains 2 marks.

What biologists are doing

Biologists 'do' lots of different things. Some work in laboratories isolating genes and attempting to introduce them into cells; others might be attempting to find simple methods to prevent elephants damaging crops. The range is huge and all are acceptable ways of approaching this section and achieving high marks. What is important is that you focus on the methods being used to answer the question or solve the problem.

The suggested word limit is between 1500 and 2000, so this section should not be a long, detailed account of complex techniques but an accurate, concise summary of the methods employed. Use your research to find the details and then summarise them in your own words.

Describing data or solutions

What you write in this section must be linked to the problem or question you have chosen. In many cases, this will be actual results; in the case of a solution it may be more descriptive. You can use any form of illustration to show the results, so it need not necessarily be a table of numbers. It is important that you describe the results carefully to show you understand how they demonstrate this might be a possible answer or solution.

Types of data or evidence you might find

(a) A simple graph

In this case the report was concerned with the question 'Can foot-and-mouth disease in cattle be controlled by developing a vaccine?' Following research into exactly how such a vaccine was developed, the graph shown in Figure 1 was included to demonstrate the effectiveness of the solution.

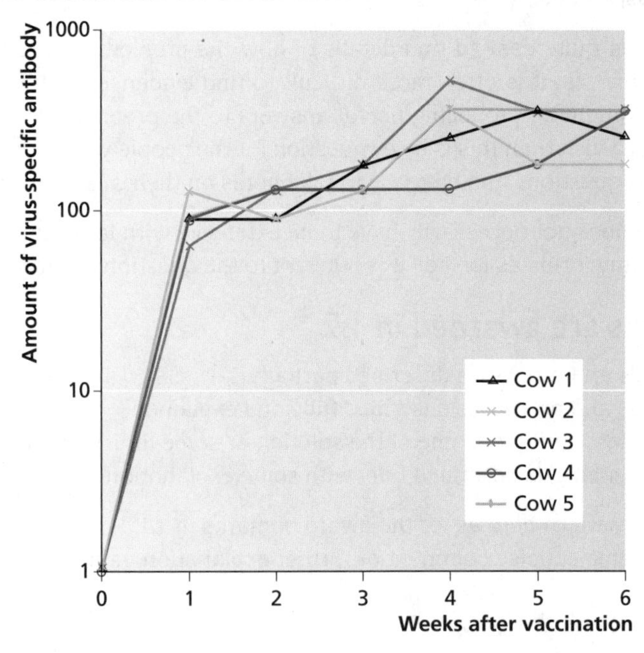

Figure 1 Graph showing how the amount of virus-specific antibody increases rapidly with time in five cows after they were vaccinated against foot-and-mouth disease virus

This is quite straightforward evidence; how this demonstrates the effectiveness of the solution can be described easily.

Knowledge check 7

Look at the data given in Figure 1. (a) What exactly do they show? (b) Why might this not indicate that the cows would not catch foot-and-mouth disease in the future?

(b) Basic information described

X-SCID is a rare but severe condition in which the immune system fails. Sufferers need to be isolated in sterile 'bubbles' or given a bone marrow transplant. The question raised was 'Can gene therapy be used to treat X-SCID syndrome?'

This is the evidence described.

'Of the 20 patients treated all have shown improvements in their immune system and 18 have been able to live normally outside a sterile environment. Unfortunately four of these have developed leukaemia-like symptoms and one has died.'

This provides evidence for further comment and also for a discussion of risks and benefits later. However, in this form it would only qualify for 1 mark.

(c) A table of data

Table 1 Numbers of cheetahs in captivity

Year	1999	2000	2001	2002	2003	2004	2005	2006
Numbers of cheetahs	1290	1315	1371	1340	1349	1382	1433	1408

The report is concerned with looking for effective methods of conserving rare species such as cheetahs using a captive breeding programme.

There is evidence in the table of the success of the programme. For the award of 2 marks, this would need further comment and explanation in terms of how the data have been collected and whether any of the cheetahs has been reintroduced into the wild.

(d) Other evidence

There is no limitation on the type of evidence that can be used and gain credit. Therefore photographs, maps, videos etc. are all useful, provided they are relevant.

1.3 Is the work of the biologists valid and reliable?

What is required in 1.3?

- You are expected to include relevant evidence in the form of graphs, tables, diagrams, illustrations etc.
- This information must be used to explain why the methods and processes you have described are valid and reliable ways of finding an answer or suitable solution to the problem.

Examiner tip

Throughout your report, use subheadings that match the criteria to ensure you address them all. It is particularly important here.

Your report should make it clear that you are considering how the data or suggested solutions can be shown to be both valid and reliable in the scientific sense. This means that you must do more than just describe them. You must explain carefully how the data or suggested solutions show that the methods were reliable and valid conclusions were drawn, or that the solutions suggested are effective.

We shall look at validity and reliability in more detail in Unit 6. Here, you should consider the following.

Reliability is about 'repeatability'. In theory, if variables are well controlled then repeats should always give the same result. Other scientists should be able to repeat the investigation and produce identical, or very similar, results. Can you find any evidence that results have been repeated or that there are sufficient data to be reliable?

Validity is about how confident you are that the conclusions drawn from the investigation are correct. Remember it is not unusual for different interpretations to be placed on similar data. So it is important to consider carefully whether the method employed actually measured what was intended, or if there might be errors that would affect the conclusion. In your explanation, consider what data or other evidence are available to justify the conclusions drawn or to show the solution was effective.

Knowledge check 8

An investigation produced data showing that rose fruits (hips) are larger in higher light intensities than in shaded areas. Why would it not be valid to conclude that this effect is due to photosynthesis?

Discussions on variability and reliability also provide good examples of what is required throughout this report. Simple statements of opinion will gain little credit unless there is **evidence** that is explained.

How marks are awarded in 1.3

1.3a Valid/reliable data or effective solutions described briefly (1 mark)

1.3a Valid/reliable data or effective solutions described in more detail and explained using graphs, diagrams, tables or photographs etc. which are well-integrated into the text (2 marks)

1.3b Brief explanation of why methods and processes are appropriate (1 mark)

1.3b More detailed explanation of why methods and processes are appropriate (2 marks)

Criterion 2: Implications of the solution

In this section, the wording of the criteria can be confusing. The main thing to remember is that here it is **all about the solutions** that have been found or suggested. It is not about the original problem that you described.

What is required in this criterion?

2.1 Identify two implications of the solutions to the biological problem or question (4 marks)

2.2 Evaluate the benefits and risks of the suggested solutions to humans, other organisms or the environment. (4 marks)

2.3 Discuss alternative views or solutions to the problem. (4 marks)

An evaluation of the advantages and disadvantages of the suggested solutions and a discussion of the possible effects on living organisms and the environment, as appropriate, is also needed.

You will see from the list above that this is an important part of the 'How Science Works' criteria.

Examiner tip

Developing your skills in 2.1 will not only help you in Unit 3 but also in other units at AS and A2.

2.1 What are the implications of the solution?

What is required in 2.1?

- You need to identify and discuss *two* implications of the solutions to the problem or question you have identified.

What is meant by implications?

There are four types of implications you might consider — ethical, social, economic and environmental.

Ethical implications

You must be clear what is meant by an ethical issue. Ethics is a wide subject area in itself. You are expected to understand both what is meant by an ethical viewpoint and that different people may hold different ethical views on the same issue. Many such issues are not simply 'right or wrong'. When commenting on ethical issues, either in your report or in a unit test, you must give a balanced reasoned argument rather than expressing a strong personal view or repeating common phrases.

Abortion: an example of an ethical issue

There are often strong debates between two conflicting ethical positions.

Anti-abortionists base their views on the principle that a human life begins at fertilisation and therefore destroying a fetus, no-matter how old, is killing.

Those who take a more liberal view argue that abortion should be allowed in some circumstances because there are other factors to take into account — for example, the rights of the mother as well as those of the fetus.

It is an interesting exercise to consider your own ethical position. This means not just taking your pick of fixed ideas but being able to explain logically why you have selected this position. A part of this would always be demonstrating that you understand alternative views but have well-argued reasons for taking your own view.

Many topics that might be chosen for Unit 3 have solutions with significant ethical issues.

Social implications

Social implications are those that will have a direct effect on other human individuals or groups.

Population control: an example of a social issue

Overpopulation is a serious issue in many countries and some, such as China, where there is a strict one-child policy, have taken strong action. This also raises ethical issues, but the social consequences can be just as important. These might include the reduction of extended families and, given the national preference for male heirs, to an increase in practices that result in severe gender imbalance.

Economic implications

In many cases it can be simply a question of expense and affordability on either a national scale or an individual scale.

The cost of new drugs: an example of an economic issue

New innovative drugs cost tens of millions of pounds to bring to the market. Drug companies need large profits to pay for this research and hence new drugs may be so costly that individuals or organisations such as the National Health Service cannot afford them. Newspapers often report cases of patients with life-threatening illnesses who are denied treatment with the very latest therapy.

In this context you might want to find out more about the body entrusted with deciding whether new treatments should be made available on the NHS. It is called NICE (The National Institute for Health and Clinical Excellence).

Knowledge check 9

Some people argue that the government should fund all science. Suggest one reason to support this view and one reason to oppose it.

Knowledge check 10

What is your opinion of the use of GM crops? Give one reason to support your opinion and one reason that another person might give for taking the opposite view.

Examiner tip

You double your mark here (and in some other sections) if you add a short discussion to explain your point, rather than just stating it.

Examiner tip

Choose a small number of the most important benefits and risks and evaluate those in greater depth.

Knowledge check 11

Fresh chicken often contains traces of dangerous *Salmonella* bacteria. To eliminate this, one suggestion has been to feed growing chickens with high doses of antibiotics. Give one advantage and one disadvantage of this treatment.

Knowledge check 12

What methods, other than antibiotic treatment, could be used to reduce *Salmonella* infections in chickens?

Environmental implications

Environmental implications are concerned with the effect of the solution on other plants, animals and ecosystems. An obvious example is the debate about GM crops. Will the benefits of increased production and reduced pesticide treatment outweigh the risks of possible gene transfer, through pollination, damaging ecosystems? You would be expected to discuss this by showing an understanding of both sides of the argument, not just stating one fixed point of view.

How marks are awarded in 2.1

2.1(a)	Identifying clearly two implications of the suggested solution	(2 marks)
2.1(b)	Further explanation of each implication	(2 marks)

2.2 Evaluating benefits and risks

What is required in 2.2?

- Identify the benefits and risks involved in the solution you describe. You have to describe them fully and then evaluate them.

What is an evaluation?

Evaluating is one of the skills that you are expected to develop at AS and A2. It means that you are expected to select some relevant information and use it as evidence in a short discussion. You must think carefully. Almost all solutions you come across will have some drawbacks — most are not perfect. An evaluation will discuss these drawbacks and come to a conclusion. This is an important skill for both Unit 3 and Unit 6.

In this context you can only achieve high marks by adding some relevant comments to support your ideas, rather than just making a list. So, do not be tempted to make a long list of everything you can think of with simple comments, as this will not provide evidence for higher marks. A consideration of the balance between benefit and risk is a good way to start evaluating, but remember, you must *explain* any conclusion you make. You might feel the benefits outweigh the risks; your explanation must include the reasons why you feel that way.

How marks are awarded in 2.2

2.2a	Benefits or advantages listed briefly	(1 mark)
2.2a	Benefits or advantages explained and discussed	(2 marks)
2.2b	Brief evaluation of some risks or disadvantages	(1 mark)
2.3b	Balanced discussion of risks or disadvantages compared to benefits	(2 marks)

2.3 What are the other options?

What is required in 2.3?

- You need to identify *two* alternative solutions to the problem you have chosen that are different from the one suggested in the research you have described.
- Both these alternatives have to be described and explained.

While scientists continue to make new discoveries and our knowledge is increased, progress is often built on painstaking work over many years — sudden spectacular advances are rare. Many of the initial discoveries reported in newspapers are over-enthusiastically hailed as major breakthroughs when the truth is that many years of research and trials will be needed to convert such discoveries into practical solutions. Gene therapy and the use of stem cells are good examples from medical research.

It is an important part of scientific research that different research groups have different ideas and theories for which they try to find evidence. There is often fierce debate about conflicting theories before sufficient evidence accumulates to support one rather than the other. Until this happens there are several alternative ideas or solutions that are equally valid.

Therefore, no matter what the subject of your report, it is likely that there is more than one solution or more than one alternative idea.

Examiner tip
Make sure your report shows you understand that suggested solutions may need further work and that other solutions may be just as acceptable.

How marks are awarded in 2.3

2.3a One alternative method or solution identified (1 mark)
2.3a One alternative method or solution described and explained (2 marks)
2.3b A second alternative method or solution identified (1 mark)
2.3b A second alternative method or solution described and explained (2 marks)

Criterion 3: Using researched information

This section is also important for Unit 6. It is a good example of how you are expected to develop your skills throughout the course. The most important point to remember is that the information you find must be reliable and accurate when judged from a scientific point of view.

What is required in this criterion?

3.1 Use information from at least three sources, including web-based and non web-based sources, in your report. (4 marks)
3.2 Provide information about the sources in a recognised scientific way and indicate where they have been used in the report. (4 marks)
3.3 Evaluate at least two references used in the report. (4 marks)

3.1 Use information obtained from three or more sources

What is required in 3.1?

- You must show you have used at least three sources in total.
- At least one must be a web-based source and one must be a non web-based source.
- Quotes from sources must be used effectively within the report and identified clearly.

The key word in this criterion is 'using'. This does not mean just copying and pasting large sections of text from sources. It means using the information they contain to explain your points accurately in your own words.

To provide evidence that you are using the sources you must place references in your report at the points where the information is used. These references must then be linked to your bibliography. You can do this by numbering your sources and placing the number in brackets in your text. It is equally acceptable to use footers to list sources that are relevant to that particular page; you will need to use a small font size to do this.

Quoting from sources

You will gain credit for doing this effectively and indicating the origin of your quote. This means the quote should form part of your text within a continuous piece of writing; it should not be just an isolated piece of information.

Here is an example from a report considering the introduction of grey wolves to control red deer populations in Scotland.

> The last wolves were killed in Scotland around 1684 [1] and their reintroduction brings mixed feelings from people living in the areas that might see wolves again. Wolves were originally hunted to extinction because they threatened farm animals and gained a fearsome reputation. Much of this may be unfounded as 'the research shows that wolf attacks on humans are rare, and fatal attacks even rarer' [2]. Obviously farmers are still unsure of the effect on their own animals, but red deer are also a major problem as their grazing competes with that of upland sheep flocks.

Examiner tip

Remember this is evidence of research skills so your school textbooks will not be accepted as a non-web source.

This simple example shows how useful information [1] was included to illustrate that wolves might be judged by reputation alone and the quote [2] gives an indication that threats to humans are unlikely.

How marks are awarded in 3.1

3.1a	One web-based source used	(1 mark)
3.1b	One non-web based source used	(1 mark)
3.1c	Evidence of use of at least three sources in total	(1 mark)
3.1d	Quotes from sources used effectively	(1 mark)

3.2 Provide information on sources and link references to text

What is required in 3.2?

- Clear and accurate referencing in an acceptable scientific form
- Evidence that you have used all your sources in your report

Examiner tip

Simply copying and pasting a web address is given little credit as an accurate description of a web-based source.

Naming sources accurately in a bibliography

A bibliography must be listed in some recognised scientific format. Scientists use a generally accepted format called the **Harvard System**.

It is easy to find simple examples of how to reference a wide range of sources by researching the Harvard System. Almost all academic institutions publish a simple freely available guide to what is required.

Here are two examples.

A web-based reference will be in the following format:

Authorship or Source, Year. *Title of web document or web page.* [type of medium] (date of update if available) Available at: include web site address/URL (Uniform Resource Locator) [Accessed date].

The copied web address would look like this:

www.nhs.uk/news/Pages/Howtoreadarticlesabouthealthandhealthcare.aspx

The correct reference would be:

NHS UK Dr Alicia White, 2009. How to read health news. [web page] Available at www.nhs.uk/news/Pages/Howtoreadarticlesabouthealthandhealthcare.aspx [accessed 25/06/2012]

A scientific journal will be in the following format:

Author, Initials, Year. Title of article, *Full Title of Journal*, Volume number (Issue/Part number), Page numbers.

A correct reference might be:

Meselson, J. and Stahl, F. W. (1958) 'The replication of DNA in *Escherichia coli*', *Proceedings of the National Academy of Sciences*, 44 (7), 671–682.

Knowledge check 13

Using the Harvard System, what is the correct format for a book reference?

Knowledge check 14

Use this format to write the correct bibliography reference for this book.

How marks are awarded in 3.2

3.2a	A few sources are referenced accurately	(1 mark)
3.2a	Almost all sources are referenced accurately in bibliography	(2 marks)
3.2b	Some references are linked to the text	(1 mark)
3.2b	All reference material is linked to the text	(2 marks)

3.3 Evaluating researched sources

What is required in 3.3?

- A review of the scientific credibility of some of the sources you have used
- A critical review of any data or evidence you have quoted

Evaluating sources

This is a key 'How Science Works' skill. It is needed for both Unit 3 and Unit 6.

For Unit 3 you have to **evaluate two sources**. Remember that evaluating means that you must come to some opinion that must be based on evidence and discussion, not on vague assertions. You must always evaluate from a scientific point of view. In this case the key question is 'would this source be regarded as reliable by groups of other scientists and why?' So you need to think like a scientist to do this.

Examiner tip
For each source select a few pieces of evidence and discuss these concisely, rather than making short comments on a long list.

What evidence might be considered in evaluating sources?

The following are points you might consider when making a judgement about the sources you find. Remember it is better to consider a few of these in detail rather than making brief comments on all of them.

- Who wrote the source? What evidence is there of their scientific credibility? Simply having BSc or a list of other qualifications does not automatically give their work credibility.
- Has the source been quoted by other scientists? You can sometimes find evidence in online journal sites to 'citations' by other scientists. This means others have used the work to explain the background to their own work or in discussing their results. It is strong evidence of acceptability.
- Has the work been peer-reviewed? All scientific investigations published in well-known journals will have been peer reviewed. This means that a senior scientist has supervised the research and the research paper has been sent to several reviewers for comments. The reviewers are experts in the same field; they check that the methods used are sound and explained clearly and that the conclusions drawn are valid, given the data presented. This process is an important part of validating new scientific information.
- Can some of the detailed information in the source be found in other reliable sources? This is called cross-referencing and is useful evidence. You must give the names of the other sources and examples of the information compared.
- Does the website or magazine contain advertising for products linked to the information it contains? If so, this might well influence the information it contains.
- Is the information linked to any pressure group? Might there be a strong reason for the information the source contains being selective? One powerful way to influence people is to select persuasive evidence, ignoring alternative views and conflicting findings.
- What is the date of publication of the source? Is it possible that more recent findings have contradicted the evidence you are using?

Examiner tip
Simply stating that a source is peer-reviewed is awarded little credit. You have to explain briefly, giving evidence that you understand what peer review means.

Evaluating data

You will find some information on this, and on validity and reliability, in section 1.3. You can be credited with marks for 3.3 if you have evaluated the data that you used in 1.2 and 1.3.

What might be considered in evaluating evidence or data?

- How large was the sample? Was it truly representative or large enough to be reliable?
- Are the data consistent? Do they show a lot of variability? Is there a standard deviation? What does this show? (You can find out more about standard deviation in the Interpreting section for Unit 6 on p. 39.)
- If you are quoting other evidence, then how was this information gathered?

See Exercise 6 (p. 60) for an example of how evidence can be critically reviewed.

Examiner tip
Evaluating the evidence or data in an objective way is often omitted or attempted poorly in reports. Make sure you include a full evaluation.

How marks are awarded in 3.3

3.3a	Evaluation of two sources	**(2 marks)**
3.3b	Evidence or data from sources investigated	**(1 mark)**
3.3b	Evidence or data investigated and evaluated	**(2 marks)**

Criterion 4: Presenting your report

Your report should be an interesting and attractive piece of writing. You can choose any type of format provided that you can meet the criteria.

What is required in this criterion?

4.1 Correct spelling, punctuation and grammar; a well set out report organised in a logical sequence (2 marks)

4.2 Technical language suitable for AS is used; visual material is presented that is relevant and helpful to understanding the report (2 marks)

4.1 Accuracy and organisation of your report

What is required in 4.1?

- You must ensure that you write clearly paying attention to spelling, particularly the spelling of technical and scientific terms.
- Your report should not contain long, purely descriptive passages.
- Your report must have a logical sequence with clear subheadings.

Using subheadings

The use of subheadings has already been recommended several times in other criteria. They have several advantages:

- Subheadings indicate whether you have a logical sequence to your report.
- The use of subheadings makes it easier to check that you have addressed all the criteria. It is useful to match your subheadings to the criteria.
- Using subheadings will prevent you from drifting into lots of description instead of doing the analysis required to meet the criteria.

How marks are awarded in 4.1

4.1a Spelling, punctuation and grammar are mainly correct and the report is reasonably well organised (1 mark)

4.1a Spelling, punctuation and grammar are correct and the report is well organised (2 marks)

4.2 Using technical language and visual material

What is required in 4.2?

- Technical language appropriate to AS
- Suitable visual presentations to make points clear and enhance the appeal of the report to the reader

What is a 'visual'?

'Visuals' can be any type of display that is suitable to the topic of your report. These might include:

- maps
- charts

- tables
- diagrams
- photographs
- graphs

There is no limit to what might be included, but it *must* be relevant.

Note that:
- enhancing the text for the reader does not mean making the report look 'pretty'.
- credit will only be given where illustrations are relevant and useful to the report; to ensure that they are useful all illustrations must be referred to in the text and their source acknowledged

Examiner tip
Use titles beginning with 'Diagram/graph etc. to show...' for illustrations. If you cannot complete the title sensibly then why is the visual included?

How marks are awarded in 4.2

4.2a Good technical language with some visual elements (1 mark)
4.2a Good use of technical language with visual elements that are used clearly within the text (2 marks)

Unit 6 Practical biology and investigative skills

Individual investigation

What is expected in Unit 6?

You are expected to produce a report of an individual practical investigation that will give you the opportunity to demonstrate your 'How Science Works' skills at a standard appropriate for A2. The recommended length of the report is between 2700 and 3300 words. There is no penalty for going beyond this but most very long reports contain a great deal of repetition and irrelevant detail.

Your report will be assessed using five main criteria as shown in the table:

Criterion	Main points	Marks available
Research and rationale (R)	Use researched information to provide a context, plan your investigation and analyse your results	11
Planning (P)	Design a safe investigation; control variables and produce meaningful data	11
Observing and recording (O)	Record data precisely and act on possible anomalies	8
Interpreting and evaluation (I)	Interpret results using statistical analysis; explain results using biological information; evaluate findings	9

Criterion	Main points	Marks available
Communicating (C)	Present the report scientifically and clearly; list sources in an accurate bibliography and evaluate them	6
Total		45
Unit 6 makes up 20% of the total A2 mark		

How Unit 6 is assessed

Unit 6 is assessed differently from Unit 3, the main difference being that each of the main criteria is assessed as a whole and not by adding up marks from small individual parts. The technical way this is done is called hierarchical marking. You do not need to know the exact details of this, but the principle is important to the way in which you present your report.

As you will see in the rest of this book, each main criterion has several subsections labelled (a), (b) and so on. Each is assessed individually and assigned to a mark range. The final mark for the whole criterion is limited to the lowest range of any subsection. In awarding a particular mark range for the whole criterion, every subsection must meet that standard. An extreme example might be a student awarded the following marks for planning: P(a) 7–9, P(b) 7–9, P(c) 0–2. In this case, the maximum mark for planning is 2 because all three subsections do not meet the requirements for a higher range. This is extremely rare, but could occur if there was no real trial investigation — hence P(c) for 3–6 or above would not be met.

This might sound unfair but it is the same for everyone and because this is A2 you are expected to be able to bring together your skills into a coherent investigation, not isolated pieces.

Examiner tip

Do not write many pages on one section then only a few lines on others that carry the same number of marks. The most common problem is a very short evaluation.

Important implications of the assessment method

- You must make sure that you address all sections of each criterion. You can lose a lot of marks by carelessly omitting criteria.
- The most sensible way to avoid omissions is to use subheadings in your report that match the criteria.
- Always review your report using a Unit 6 checklist. Edexcel publish a detailed student checklist; there is an abbreviated version at the end of this book (see p. 68) which covers the most frequent mistakes.

Getting started

Before you start make sure that you are thinking at the right level. To achieve high marks at A2 you need to have moved on from AS and show you are capable of more mature scientific thinking. This might sound rather vague but asking yourself some simple questions might help:

- When you visit a new website do you question its origin or purpose?
- If you read about some new 'superfood' do you ask what evidence there is to support the claims?
- When you carry out a core practical do you think about the science behind the technique or whether what is being measured is scientifically meaningful?

- Do you understand that the material in an A-level textbook may be only a simplified version of what is really happening?

Examiner tip
All the criteria are assessed on the *quality* of what you do, not as an exercise in simply 'done that'.

Demonstrating a scientifically objective view becomes particularly important in Unit 6 when you are asked to provide a detailed biological background or evaluate your findings. Therefore you must start with a determination to investigate an interesting question, not spend your time looking for lots of things to copy. Good scientists are always asking 'How do we know that?' and 'What evidence is there for that?'

What to investigate

This is a question that you must discuss with your teachers. You have a wide choice, as long as it is practical biology at A2 standard. However, different schools and colleges may choose to organise this in different ways.

There are, however, some important things to avoid:

- Do not copy a core practical with a minor variation. The examiners expect you to have completed these during your course, and if you already have detailed instructions on important features, they cannot give you credit as evidence of your own skills.
- Do not attempt to investigate hypotheses that do not have a biological background. For example, the cost of two different brands of toothpaste is a question about marketing and brand image, not about A2 science.
- Do not investigate hypotheses that are about psychology and not about biology. This is particularly true if you are also studying A-level psychology. Remember you are not allowed to submit a piece of coursework that is the same or very similar for two subjects.
- Do not choose an investigation just because your friends are doing it. If many parts of your investigation are the same or similar to others, then an examiner will not be able to identify your contribution. If investigations are too similar they may be rejected.
- Do not choose an investigation that does not involve much practical work. Basic questionnaires and simple information collecting will gain little credit or could be rejected.
- Avoid copying investigations you may have used at GCSE or AS, such as basic enzyme investigations where many of the methods have significant flaws or are demonstrations rather than investigations.

Reading this book implies that you have chosen to study biology. Congratulations! You now have a great opportunity to investigate some interesting questions about the world around you. Don't waste it by attempting to demonstrate some well-documented 'fact'. It is quite easy to produce unique data without the use of complex equipment or highly sophisticated techniques.

A simple example might be an investigation into feeding preferences or colour perception in garden birds. It is easy to make up a simple medium from bread dough and form it into shapes that could be coloured with food dyes. This would provide opportunities to form different hypotheses.

What makes even this simple idea useful is that it isn't quite that simple! You would need to think carefully about variables and carry out trials to make your method reliable. There is also a range of information to research.

You could choose to use a standard technique — this is not a problem. Just make sure that what you are investigating has an interesting question that gives you the opportunity to show some planning skills by using the technique in your own way.

Research and rationale

What is expected in this criterion?

- Evidence of research into the hypothesis to be investigated that is referenced clearly within the text
- The relevant biological background to the investigation is explained
- Some explanation of why this might be of interest to biologists
- Researched information is used in planning and in explaining the data collected

As you can see, evidence for research and rationale must be included in several parts of your report.

Research

In this section it is assumed that you have developed basic research skills throughout your AS course. You are strongly recommended to go back and read the section on 'researching information' in Unit 3 (pp. 19–21). Exactly the same rules apply here.

Examiner tip
Examiners are looking for concise, accurate explanations of the background to your investigation. Make sure that you focus on your hypothesis and avoid long theoretical accounts that include irrelevant detail.

Biological background

The key word here is **relevant**. This is also a test of your ability to present information in a logical sequence. Always keep in mind exactly what you are investigating and do not be tempted to include lots of information that has only limited value in explaining the background to your investigation. Many candidates include numerous pages of copied material without explaining exactly what they are investigating.

Most of this must be in your own words. You can use relevant quotes and some diagrams but long sections of copied information will gain no credit. To achieve higher marks you must explain in some detail, not just at a superficial level.

Rationale

A **rationale** means explaining how your investigation might be biologically interesting or relevant. Your rationale ought to be phrased in terms of why this investigation might be interesting to other biologists. Above all, it must be scientifically sound. This does not mean that you need to write a long justification — you just need to show some understanding. A good example is given in Exercise 8 where the ability of ivy to change the size of new leaves depending upon light intensity is part of a wider explanation of this plant and its ecological niche.

You are also expected to use your scientific judgement. It is easy to find many references to the antibacterial properties of garlic. However, suggesting that this might be the answer to the worldwide problem of antibiotic resistance in bacteria is likely to be going too far. A little more detailed research would show that the active ingredient in garlic is broken down by heating and the dosage required to produce

significant effects is equivalent to several whole raw garlic bulbs per day. This might have significant social effects, thus limiting its usefulness!

The advice therefore is to research carefully and think about the biology in a wider sense.

Planning

What is expected in this criterion?

- You should demonstrate that you have a clear plan of action for your investigation. P(a)
- You should use trials to decide on some important features of your plan rather than making arbitrary choices. P(c)
- You should design your method to control as many variables as possible. P(a)
- Your main dependent variable should be measured as precisely and reliably as possible. P(a)
- All important risks should be assessed and minimised to produce a safe procedure. P(b)

Knowledge check 15

What is meant by (a) the dependent variable and (b) the independent variable?

The words 'as possible' mean that you will be judged against what can reasonably be expected of you as an A-level student in a school/college situation.

This criterion is one of the most important pieces of evidence of your investigative skills.

P(a) Variables

You cannot be expected to control every single variable that could possibly affect your results. However you are expected to make sure that all those that could have a significant influence are controlled. Where, for example in fieldwork, control is not possible then monitoring some selected abiotic factors could be important to confirm that they did not affect the data you collected.

First, you must concentrate on your main chosen dependent variable. How is this to be measured accurately and reliably? Then look carefully at your chosen independent variable.

This important area is where it will show whether you are thinking for yourself or simply following instructions. Even many procedures in books are not thought out carefully.

Variables in the field

(a) Light intensity is notoriously difficult to measure because it changes continuously throughout the day and according to the weather. If you have chosen this as your main independent variable then you must think of possible ways to make the most accurate assessment. This obviously won't be simply holding a light meter at arm's length for a few seconds. You might have time restrictions at your location, so think of different ways of taking an average reading. Look around and check what might happen at different times of day or in different seasons and so on. This is a great opportunity to trial some different methods and find which might be more reliable. There is no simple 'right' answer but you will gain credit for trying out some practical solutions and selecting the one with the fewest drawbacks.

(b) There are too many other abiotic measurements to go through here but they do provide good opportunities to think about standardising your method, particularly if they are going to have a big effect on your data. At what depth is the water temperature to be taken? How deep is the soil sample? Attention to details such as these will provide evidence that you are thinking carefully about collecting reliable data.

(c) Just because something is not on your initial plan there is no excuse for not using careful observation and changing things accordingly:

- Imagine you are taking a transect across a rocky shore and come across a small stream running down to the sea. Do you carry on regardless? You are sampling molluscs and find some are in crevices and some on open rock. What should you do?
- Imagine you are investigating the distribution of a single plant species using random sampling with a quadrat. You quickly realise that other species are obviously having an effect. What should you do?

Examiner tip

There is common misunderstanding about species diversity and species richness. Make sure that you are accurate if you intend using these terms in an ecological investigation.

There are no ready-made answers to these questions — it depends on the details of your hypothesis and on the location. What is expected of you is some sound scientific 'common sense'. One approach might be to amend your sampling to make sure that you do not introduce another independent variable. Another might be to continue, but to record carefully which measurements are which and analyse your data more fully. Different solutions might each have their merits and provide evidence for high marks. What would indicate poor planning would be to introduce this problem as a limitation later, therefore suggesting that your data and conclusions are meaningless.

Knowledge check 16

What is the difference between species diversity and species richness?

Variables in the laboratory

Some substantial flaws can arise from failing to consider concentrations correctly in SI units. Concentrations in per cent are much easier to make up as 1% is simply 1 g made up to 100 cm^3 of solution. However, there are good reasons for using the correct units but you need to use common sense at times. For example, it is not sensible to record time in seconds if the time scale is several hours.

The main SI units

The table shows the main SI units used in A-level biology.

Measurement	Unit	Correct label
Time	Seconds Minutes Hours	s min h
Length	Metre Millimetre Micrometre Nanometre	m mm (10^{-3} m) μm (10^{-6} m) nm (10^{-9} m)
Volume	Decimetre cubed Centimetre cubed	dm^3 cm^3
Amount of a substance	Mole	mol
Concentration	Moles per decimetre cubed	mol dm^{-3}
Temperature	Degrees Celsius	°C

Knowledge check 17

What is the SI unit of concentration?

Knowledge check 18

Why does a 1% solution of glucose not have the same concentration as a 1% solution of sucrose?

Examiner tip

You are in a much stronger position to control things in a laboratory, so more attention to detail will be needed.

Examiner tip

Vague statements such as 'room temperature' or 'left on a windowsill' will gain little credit.

Concentrations can raise problems:

- Trying to repeat the core practical on antibacterial properties of plant extracts by adding substances for comparison is meaningless unless you know the concentration of each.
- Individuals of different mass do not have the same concentration of caffeine in their tissues even if they take the same dose.

Sample numbers

It is not possible to give fixed rules about sample size. This is because each type of investigation is different. It might take a good deal of effort and care to collect data from ten samples where lots of preparation is involved, but this would be poor for an investigation just measuring a simple dependent variable. Whatever the investigation, you must have sufficient data to be able to make a meaningful scientific conclusion within any limitations there might be.

Where data are readily available, then a **running mean** might be used. This is a way of confirming that there is a stable mean before statistical analysis and selecting sample size. Starting with the first reading a graph is plotted of the mean against number of samples. At first, with few samples, the mean varies a great deal with each additional reading; it eventually settles to a consistent value. At this point sampling may stop, or in some cases this number of samples is doubled to make sure the mean is stable enough to be tested. You are unlikely to be able to do this unless you have a lot of readily available data.

An example of a running mean is shown in Figure 2.

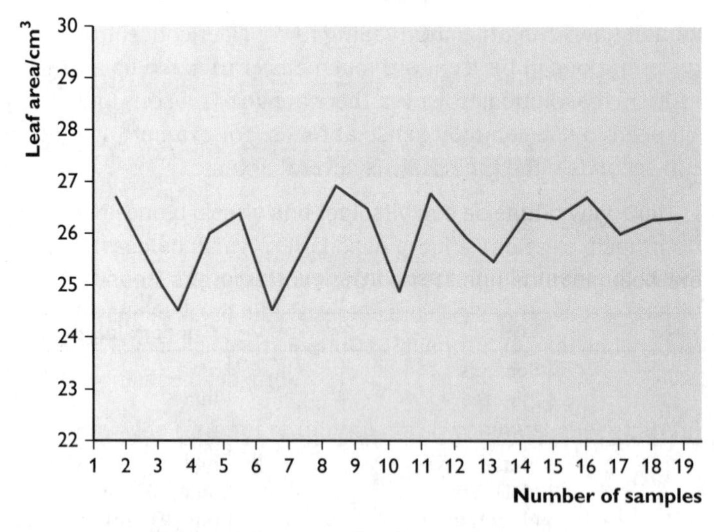

Figure 2 An example of a running mean

Knowledge check 19

Look at the graph of the running mean. How many samples are needed to establish a stable mean for these data?

If you are unlikely to be able to compile a running mean to confirm sample size, then think about how to analyse your data statistically.

Random sampling

In all investigations you need to consider how to select your samples. Even when you are using a repeated laboratory analysis, you still have to consider how to standardise the initial selection of material.

Random sampling is much more important in ecological studies where you cannot control variables as you can in the laboratory.

This section reviews some important ideas that you are expected to have studied in detail in Unit 4 Topic 5.

- It is impossible to measure everything in a whole area such as a shore or a long stream or river. Therefore, samples of smaller areas are taken, from which inferences can be drawn. This is useful because it makes investigations possible, but might not tell the whole story. Sampling can also introduce bias. If you are measuring the heights of some plants in an area you are much more likely to measure taller plants than to search for shorter examples. If you sample coloured molluscs on the shore you are much more likely to select the brightly coloured examples.
- To avoid bias, random selection of samples is used.
- The method of random sampling you choose should be matched to the habitat and organism you are investigating.

Examiner tip

Always consider your hypothesis, and exactly how you intend to analyse your data to accept or reject it, at the planning stage.

Sampling on a random grid

The principle here is simple. A rectangle is set out which has a vertical and horizontal axis usually made up of measuring tapes. A random number table is used to select pairs of numbers that are used to mark a line on each. Where the two lines meet is the random sampling site.

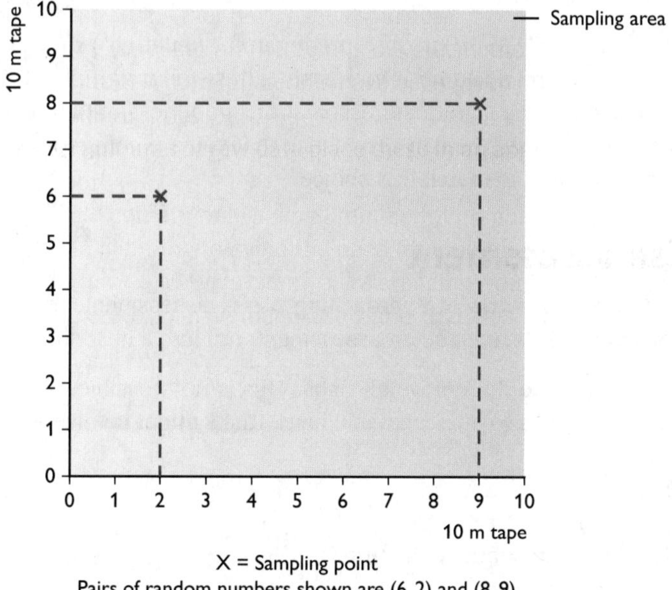

Figure 3 Grid sampling

Sampling using a transect

Here the selection of sampling begins with lying a tape across one area of the site. Random numbers are selected to indicate distance along the tape, which defines the sampling site. This is a simple **line transect**. Assessing a small area at each site along the tape using quadrats is a **belt transect**. If the distance to be tested is small, there could be a continuous belt along the whole tape (Figure 4). If the line is very long then samples could be taken at intervals forming an **interrupted belt transect** (Figure 5).

Knowledge check 20

In ecology, when might it be more appropriate to use a transect rather than a random grid for sampling?

Knowledge check 21

What differences might be expected when assessing abundance using a point frame compared to an open quadrat?

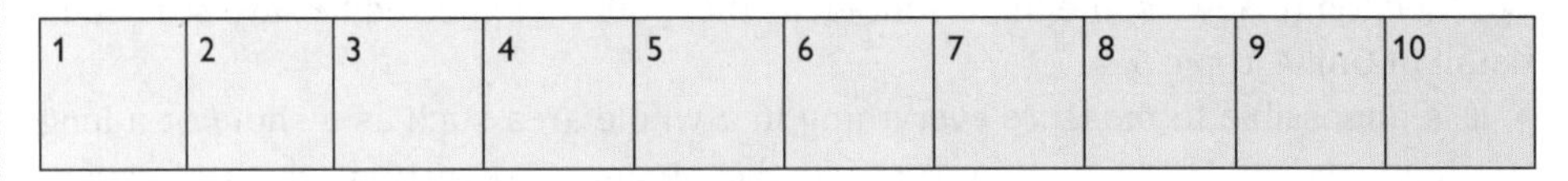

Figure 4 Belt transect

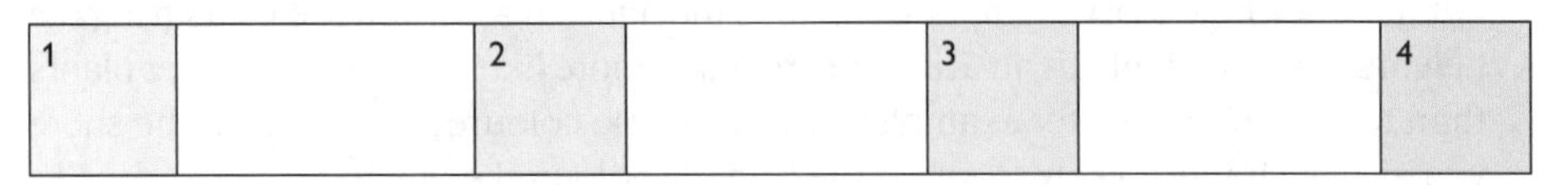

Figure 5 Interrupted belt transect

Using your knowledge from Unit 4 could provide you with opportunities to develop meaningful trials as described in the next section.

Link your plan to analysis

An important part of your plan must be to think ahead and consider exactly how you intend to analyse your data statistically. It is not uncommon for students to rush ahead with data collection only to find later that there is no logical way of carrying out an analysis.

The statistics section (pp. 36–41) in 'Interpreting and evaluating' provides much more information, but start by deciding if you wish to test for a significant difference, a significant correlation or a significant association/'goodness of fit' and then choose a particular statistical test. You can then check that you are going to collect the right type and amount of data to match this choice.

P(b) Risk assessment

There are a number of ways of approaching a risk assessment, but whatever the approach your comments must be an assessment, not just a description.

In an ideal world you would remove all risks. This is not possible, so you must take measures to reduce risks to an acceptable level. Risks might involve:

- chemical compounds
- equipment
- dangers present in outdoor sites
- possible risks to participants in testing
- protection of personal data

It is assumed that by this stage you are competent to follow basic laboratory rules such as tying hair back and keeping bags stored safely, so points such as these are given little credit.

The level of risk depends on:

- the possible consequences
- the likelihood of it happening

These factors determine the action that you might take. For example, you have probably grown bacteria on an agar plate in an incubator. If the incubator is set too high at 37°C then the risk assessment would be:

- Possible consequences — culturing human pathogens that could cause serious illness
- Likelihood — high in a school laboratory with inexperienced investigators

Both the possible consequences and the likelihood are much too high to proceed, so the rule in school laboratories is not to incubate above 30°C. At lower temperatures there is still a risk, so other standard microbiological precautions apply.

Examiner tip

Whichever risk assessment approach you use make sure that you show that you are assessing, not just listing.

Several field centres use a chart where the risks of certain activities are given a score from 1–5 for both severity and likelihood. The total is then checked and any activity above a certain level is subject to sensible restrictions to lower the risk.

Materials such as Hazcards are useful sources of information but copying, rather than using, this information will gain little credit.

You may be carrying out an investigation that has low risk. There is no need to make up a pointless trivial list just to tick this box. If you feel that risks are low then explain your reasons for this decision and, provided you have not overlooked anything, this will be accepted as correct.

P(c) Trial investigations

This is the most important piece of evidence that you need to provide to show that you are really thinking about your investigation and not simply following instructions. You cannot be awarded more than 6 marks if your trial is weak and does not lead to decisions about your method. To be awarded 7–9 marks or more you have to show that you have tested some important features of the suggested method and that your data will be as reliable and valid as possible. You cannot trial everything, but it is expected that you will look at several features of your method.

What are the features of a good trial?

- The trial concentrates on important features of the method such as the principal variables
- The trial produces some data (though these do not need to be extensive) as evidence on which to base decisions
- The trial leads to some sensible amendments to the method or confirms some choices

Examiner tip

Make sure your trial helps you to decide something important, such as how best to measure the main dependent variable or manipulate the independent variable accurately.

What are the features of a weak trial?

- The trial is initial data collection with little discussion
- The trial attempts to test a feature that is obvious — for example, that a vernier calliper is more accurate than a 30 cm ruler
- Amendments to the method do not follow logically from the trial data

Achieving the highest mark range

The highest mark range (10–11) is awarded when you have met all the criteria to a high standard and have also shown some personal ingenuity in planning. This means that you might have adapted a basic technique in an individual way or that you have thought of a different way to control or monitor an important variable. This does not need to be complex or a long account, just evidence that you are thinking individually in a scientific way.

Observing

What is expected in this criterion?

- All of your data are recorded clearly. O(a)
- Your data show that you have a suitable range of values to make a meaningful conclusion. O(a)
- Your data are recorded to an appropriate level of precision. O(a)
- Any anomalies in your data are noted, investigated and any action taken explained. O(b)

O(a) Precision of data

There are several things to take into account when considering the appropriate level of precision for your measurements:

(1) What did you use to make the measurement? A standard thermometer can be read to 1°C and if you are very careful perhaps to 0.5°C; anything more accurate would be extremely optimistic.

(2) What is the scale of measurements? You may be measuring the height of grass on a sand dune where the dimensions might be as much as 50 cm or you might be measuring the width of a thin leaf using a micrometer. A suitable level of precision would be different in each case.

(3) What is the scale of measurement and how much does it vary? A digital light meter might give a reading to 0.1 lux but if it is varying by 10 or 20 lux every time you try to take a reading then recording the intensity to 0.1 lux would not be appropriate.

There is no fixed rule about the level of precision, only that it is appropriate to what you are doing and the measuring device you are using. It is an important factor to consider in your planning.

What is important is **consistency**. If you have decided on an appropriate level of precision and this is represented by the number of decimal places in your figures then this should be the same for all your data. For example, if you decide that two significant figures is an appropriate level of precision, there is no justification for then recording means to three or four significant figures — carrying out some addition and division does not make your data more precise. Similarly, if you use a spreadsheet such as Excel to carry out some calculations, it is your responsibility to adjust the settings so that you do not print out numbers with many significant figures that are totally unjustified.

Knowledge check 22

A student records some data as 2 mm in one part of a table and as 2.0 mm in another part. Explain why these values are not the same.

O(b) Anomalies

You cannot achieve more than 3–6 marks in observing unless you review your data and check for **anomalies**. To be awarded 7–8 marks, you must identify any anomalies and explain what action you have taken. If there are no anomalies you must explain briefly why you have come to this decision. If you decide to include all your data in analysis you must explain why you chose to do this.

An **anomaly** can be any piece of data you have collected that does not fit into the pattern shown by the other results you have recorded.

You should check for anomalies as you collect data so that you can make any amendments necessary to ensure the rest of your data are reliable.

When is an anomaly not an anomaly?

You are likely to be investigating living things that may show large variations. You need to think carefully before labelling any piece of data an anomaly. In particular, check the range of the data you have collected. An anomaly is not just the largest or smallest of that range. It must show an obvious difference from other data.

Anomalies are not errors or 'wrong', unless you have made a careless mistake when counting or measuring. There may be a reason for their presence. If possible, you should investigate by looking again at the reading or repeating it.

A piece of data is not anomalous just because it shows something that you did not predict. It is whether one or more readings differ from the rest regardless of what the others show.

Detecting anomalies

There are ways of checking statistically for 'outliers' in a data set. These are beyond what would normally be expected at A2, but you can investigate them if you wish.

The best way to make a sensible judgement is to plot a scattergraph. This will give you a visual indication of any data that seem to be outside the main trend. This is a subjective decision, but if your decision is reasonable then it will be accepted provided that you explain your reasoning and any action you take.

The most important source of information about anomalies will be the repeats of your observations. In theory, if everything is perfectly under control then all repeats will be identical. In practice, this is extremely unlikely and close examination of your data might identify any that are unusual. If you do this as you carry out your investigation then you could have the opportunity to check any results that appear to be anomalous. When deciding whether something is anomalous you should take into account what you are measuring. For example, variability for a natural population measured in the field will be very different from that of a tightly controlled set of repeats in the laboratory.

Examiner tip

Use a spreadsheet to enter your data as you collect it. This will allow you to check a graph of the trends and makes it much easier to recognise anomalies.

You must take great care not to be influenced by what you expect to happen rather than what your data show.

It is likely that at this point you will want to draw a graph to aid your analysis. Details of graphical presentation are in the section on communicating (see pp. 47–51).

Interpreting and evaluating

What is expected in this criterion?

- A recognised statistical test to analyse your data. I(a)
- Conclusions from your test show understanding of a null hypothesis and 5% confidence limits. I(a)
- An interpretation of your results using biological principles and researched information. I(b)
- An evaluation of your results showing that you recognise limitations of your procedure and their effects on the validity of your overall conclusions. I(c)

Examiner tip
Always consider your statistical test at the planning stage. Use this to decide on the number and type of data. Don't try to collect data then fit it to a test.

I(a) Statistics

Even if you are not a confident mathematician, there is no reason why you should not understand the **principles** of statistical analysis without getting concerned about the mathematical theory behind the tests you might meet. These principles are covered in the following sections.

Why use statistics?

In your biology course you have probably come across investigations and data from which you have been asked to draw conclusions. Although you will have had to apply your biological knowledge, these conclusions will have been based largely on your opinion. The trouble with opinions is that everyone has their own and this is not good enough if there is to be reliable scientific progress.

Knowledge check 23
What is the probability of rolling a six using a single throw of a dice?

Many investigations collect data to discover the answer to questions such as:

- 'Is there a difference between these two sets of results?'
- 'Is there a correlation between these two variables?'

Knowledge check 24
I have just rolled two sixes in my first two throws. What is the probability that I will roll another six on my next throw?

The data collected may not make the answers to these questions obvious. So we need some rules on how we are to decide. These rules must be agreed before the data are collected, not selected later, as this would make it tempting to choose the rules to fit our ideas and they must be recognised by all other scientists.

The basic rule is about chance or **probability**.

Probabilities can be written in several ways (Table 2). In science, you are likely to meet them in decimal format.

Table 2 Probabilities

	Fraction	Percentage	Decimal fraction
Probability of tossing a fair coin and it falling as heads	½	50	0.5

A probability of 0 indicates zero probability and 1 or 100% indicates total certainty.

Examiner tip
Always make sure you use the word significant if your statistical test demonstrates this. 'Different' and 'significantly different' are not the same thing.

Imagine that you are investigating whether there is a difference between the height of plants in two different areas. You measure the heights of random samples in each area. But is there a difference?

The problem is that the height of the plants will vary in each area, so it could be possible that more, smaller plants than average were measured in one area and more slightly taller plants than average were measured in the other. Therefore, we might think that the plants in the two areas had different heights when in fact they were the same.

Most of the statistical tests you might choose will allow you to calculate the probability that your results could occur purely by chance.

The rule that scientists apply to the type of data you are likely to collect is that there must be less than a probability of 0.05 (5% or 5 chances in every hundred) that your results could arise simply by random sampling in the two areas. This is called the **5% significance level**. If you can show that the probability is less than this then you are entitled to claim that there is a **significant difference** between your two sets of data.

Exactly the same argument is true for tests for a **significant correlation**.

Null hypotheses

Testing a hypothesis using the 5% significance level requires several steps. In most cases this will be done by following the instructions for your chosen test, but you do need to understand what is meant by a **null hypothesis**. The example of plant heights given above is used to illustrate the basic steps in hypothesis testing:

(1) Start by assuming that there is no difference in the height of plants in both areas. (This is the null hypothesis.)
(2) Measure the height of sample plants from both areas.
(3) Use the statistical test to find the probability of getting the results that you have measured if there was no difference in heights.
(4) If the probability of getting results like yours if there was no difference in heights is very low (less than $p = 0.05$), then you can reject your idea (null hypothesis) and accept that there is a significant difference (the alternative hypothesis).

A null hypothesis is usually given the symbol H_0 and the alternative hypothesis the symbol H_1.

Types of statistical test

Three types of test will cover almost all the investigations you might undertake. **Correlation** should be linked to the table of statistical tests (Table 3) as should **ordinal-level measurements**.

Table 3 Statistical tests

Type of test	Common tests	Notes
Testing for a significant difference	*t*-test Mann–Whitney U test	Only for normally distributed data at the interval level only Can be used for different types of data at the ordinal or interval level
Testing for a significant correlation	Spearman's rank test Analysis of variance (ANOVA)	Simple to apply and understand Complex to apply and understand
Testing for an association or 'goodness of fit'	Chi-squared test	Only for categorical data

Knowledge check 25

A student is testing the idea that caffeine reduces reaction time.

What would be a suitable null hypothesis for this investigation?

Knowledge check 26

Rewrite the following null hypothesis in the correct format and state the alternative hypothesis.

'The volume of limpets on a sheltered shore and on an exposed shore will be the same.'

A **correlation** is a relationship between two variables, where it can be shown that as one increases so does the other (positive correlation) or as one increases the other decreases (negative correlation)

Ordinal-level measurements are measurements where it is possible to assess that one is larger or smaller than another but it is not possible to measure the actual difference between them. The ACFOR scale that you may meet in ecology is a common example. This often applies to variables that are difficult to measure and so a subjective assessment is necessary. For example, it may be possible to make a clear distinction between well-camouflaged organisms and not-so-well camouflaged organisms, but be impossible to measure.

Examiner tip
You can use a computer to perform the calculations but you must show how the data are to be set out for analysis and interpret the calculated value in your own words. A standard statement printed out from a computer programme will gain little credit.

Categorical data measurements are counts of numbers that fall into distinct groups or categories such as round/wrinkled, has/has not, blue eye/brown eye etc.

Interval-level measurements cover almost all the data collected by measuring something. You can always quantify differences between interval-level measurements such as volume of gas, time taken, length etc.

Examiner tip
Take care to use the word correlation correctly. Using it inaccurately often reveals a poor understanding of what is being tested.

You are not expected to know the more technical details of all the tests. However, do make sure you understand the principles and avoid basic errors when making your choice. Some examples are given below.

- An explanation of a normal distribution is given in the next section.
- ANOVA testing is beyond what is expected at this level. It is useful when trying to test the effect of several independent variables on one dependent variable. This is best avoided in selecting your hypothesis and planning your investigation. You will need to seek help from your teacher if you need to use this test.
- Chi-squared tests are often misused. They can only be used for **categorical data**. Investigations where this test is applicable are rare at this level, so check carefully if you are thinking about using this test. You will be familiar with categorical data from genetics — for example 'red-eye' and 'white-eye' are typical categorical counts in *Drosophila* investigations. In such cases you may form a hypothesis that there will be a fixed ratio such as 3:1 in the results. It is rare for the ratio to be exactly 3:1, so you can use a chi-squared test with a 5% confidence limit to test the 'goodness of fit' of your hypothesis. Are your data close enough to this predicted ratio?
- An association might be tested. For example, if you formed a hypothesis that rose bushes are more likely to suffer from black spot disease in rural areas than urban areas, then this could be an association. If the independent variables were rural and urban and the dependent variables were 'has black spot' and 'does not have black spot' then these would be categorical measurements. If you decided to measure the area of sample rose leaves affected by black spot then you would not have categorical data so would need to use one of the tests for a significant difference.

Accepting or rejecting a null hypothesis

For each statistical test you will need to calculate a test statistic. In most tests you will use a formula to do this. You will need to consult your teacher on how to apply these formulae. You need only do this for the test you have decided to use.

Following this calculation, you need to know the critical value of your chosen test statistic that matches your chosen level of confidence (0.05 or 5%). This involves a lot of calculations but fortunately these have been carried out and the results published in tables. If you do not have access to copies of these they are freely available by typing the name of your test and 'table' into an internet search. Some are more complicated than others so if you find one too complex, look up another. How to interpret the table is slightly different for each test (Table 4).

Table 4 Interpreting common statistical tests

Name of test	Name of test statistic	Rejecting null hypothesis
t-test	*t* value	Reject if your *t* value is higher than the critical value
Mann–Whitney *U* test	U_1 and U_2 values	Reject if the **lowest** *U* value is **equal** to or **less than** the critical value
Spearman's rank test	r_s Spearman's correlation coefficient	Reject if r_s is greater than or equal to the critical value
Chi-squared test	χ^2 (chi-squared)	Reject if value of χ^2 is greater than the critical value

Descriptive statistics

As the name implies descriptive statistics are used to describe the data you collect. Several important descriptive statistics are part of the specification so can be tested in unit tests. An understanding of their meaning is important when describing data and when evaluating your investigation.

Normal distribution

Living things often show variation. It is a vital element for natural selection to operate. Measuring one feature of a large sample of individuals often produces the pattern shown in Figure 6.

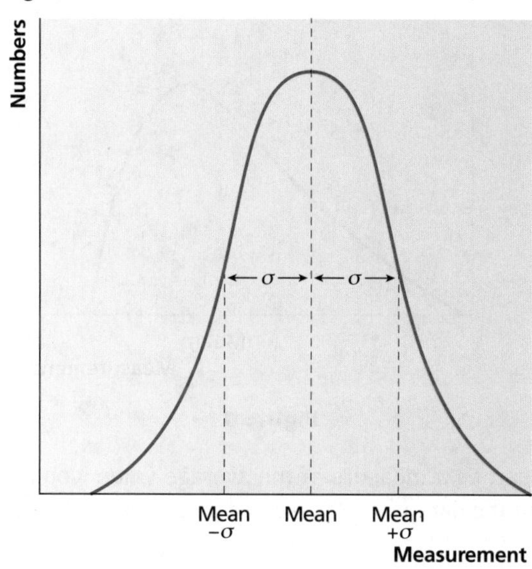

Figure 6 A normal distribution. The mean is the centre of the distribution. The standard deviation is shown as σ

In simple terms, this shows that the 'average' measurement is the one shown by most individuals and that the distribution is a symmetrical bell-shaped curve. This is what is meant by a 'normal' distribution.

Mean — the average value of a set of measurements is calculated by adding all the measurements and dividing by the number of measurements.

The word 'average' is used in lots of different ways so the term **mean** is better as it has a clearly defined scientific meaning.

If you look carefully at Figure 6 you will see that it tells us something else about the data — how much the data are spread out. If all the data are in a narrow range then the curve will be tall and thin; if the data are well spread out then the curve will be much wider and flatter. This is an important feature when trying to decide if two populations are different and can be calculated precisely. The calculation gives us a **population standard deviation**. This is given the sign σ (sigma). In a normal distribution 68% of the values lie within ± one standard deviation of the mean; 98% of the values lie between ± two standard deviations from the mean.

Population standard deviation is a measure of how much the sample data are spread out from the mean value.

Skewed data

Not every set of measurements you might take will be spread out like a normal distribution — there could be a lot more readings at one end of the scale. For example, if you were investigating the percentage cover of heather on a moorland you would soon find that it is very common and most of your measurements would be in the 80–100% range. This would not give a symmetrical curve. It would look more like the graph shown in Figure 7.

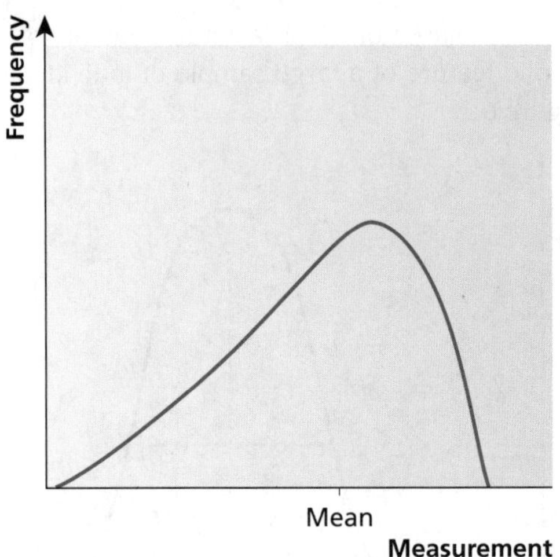

Figure 7

Median — a measure of an 'average' value for skewed data using the middle value of the range of readings, rather than the mean.

Mode — a measure of an 'average' value of data found by identifying the measurement that occurs most often in a set of results.

In this case the mean as a measure of the average value would be misleading and would not represent the data well. For skewed data such as these, the **median** value is more useful.

Another way of expressing an 'average' value is to use the **mode** or modal value of the data. For data with a normal distribution, the mean, median and mode values are likely to be the same but for skewed data they could all be different.

Error bars

Error bars are not well named as they do not show errors. They are used to show the variability of data on graphs, to enable you to analyse the data in an objective way. For example, if you measure a natural feature such as the area of leaves you would expect your data to vary over a range of values. If the error bars show a lot of variability in a laboratory investigation in which you intended to control the variables carefully, then they may indicate a flaw in your planning.

To calculate the lengths of error bars you would normally calculate the mean of your data and then the standard deviation (to be strictly accurate this ought to be standard error but this is a little beyond A-level). Error bars are illustrated in Figure 8 using data comparing the length of rats' tails in the wild with those kept as pets.

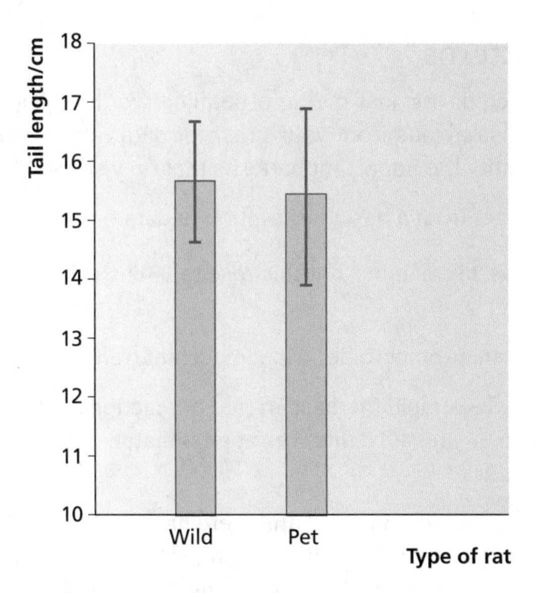

Figure 8 Presenting descriptive statistics using error bars. The mean and standard deviation of the tail lengths of wild and pet rats are shown. The error bars rise one standard deviation above and one below the mean.

A much simpler way is to use a range bar, where the bar represents the highest and lowest values above and below the mean.

A simple way of showing variation in skewed data is by using a 'box-and-whisker' plot as described on p. 51.

This type of evidence will be useful in evaluating the overall conclusions later.

I(b) Interpreting and explaining your data

Watch your language

At this point in your report you need to think about what it means to be scientifically objective. You must start by asking what your data show, not what you think they should show or what you have previously decided is bound to be the explanation. Despite using statistics, it is unlikely that you will have 'proved' anything. Scientific papers are notable for their cautious statements and lack of definite assertions.

- Don't jump to conclusions you cannot support with evidence.
- Concentrate on your own data, not what textbooks suggest should happen.
- Use cautious language such as 'supports the idea that' or 'might indicate'.
- Avoid any reference to a wrong result. Unless you made some large errors your data are what your investigation showed.
- Biology is about living things and living things show great variability.
- At A2 you do not have the full details of most of the biological processes involved. Your research will, hopefully, have begun to show you that most topics are more complicated and detailed. So take care to show you understand the limitations of simple explanations.

Trends and patterns

You will gain little credit for just giving a detailed word description of every part of your data which is obvious from your graph or data tables. You are expected to understand and identify the important patterns that they show.

This is a typical extract from a basic description of data:

At 30°C the rate was 25 $cm^3 min^{-1}$ but this rose to 33 $cm^3 min^{-1}$ at 35°C and then to 38 $cm^3 min^{-1}$ at 40°C.

Compare this with an attempt to describe important trends:

'Below 40°C there was a rapid increase in rate of reaction as the temperature increased. Between 30 and 40°C this was approximately 1.3 $cm^3 min^{-1}$ for each 1°C rise in temperature'.

Here the data have been summarised and there has been some attempt to manipulate the figures to provide further useful information. This would meet the requirements for identifying trends and patterns.

It is also important to note that this is exactly how mark schemes are constructed for data-response questions in the other unit tests.

Statistics have their limitations

We have seen in the previous section that statistical analysis provides some specific conclusions. In most investigations this is either 'there is a significant difference between...' or 'there is a significant correlation between...'. This is often important, but it is vital to remember that this is all that they show.

Examiner tip

To make sure you gain full credit for R(b) and I(b) you should include some references when explaining the data in the interpreting section of your report.

Statistical tests do not tell us anything about the reasons why there might be this difference or correlation. This means that when you attempt to explain the biological reasons why there might be this difference or correlation or the significance of it, you are just suggesting possible explanations. If you wish to achieve high marks then you must support such suggestions with reasoned arguments using the information you have researched. The language you use must show that you understand exactly what can be concluded from your data and what is just a possible explanation.

Don't forget the data

A statistical test does not take account of important patterns in data. This is particularly true of correlations.

Figure 9 shows the results of an investigation of the effect of enzyme concentration on the rate of reaction. The hypothesis suggested that there was a positive correlation between enzyme concentration and rate of activity.

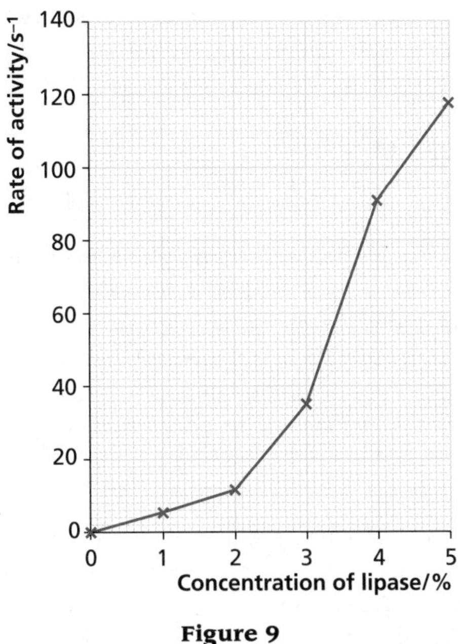

Figure 9

A Spearman's rank correlation test for these data gave a correlation coefficient of +0.94. The critical value at the 5% confidence level for six pairs of measurements is 0.89. The calculated value exceeds this critical value so we can say there is a significant correlation.

However, look closely at the graph. This is clearly not the whole story. There is not a perfect correlation across the whole concentration range and the effect at the lowest and highest concentrations should lead you to think about the science behind this. For the higher mark ranges you are expected to recognise this and to attempt to use your knowledge to explain what could be happening.

A similar problem can arise when testing for significant differences. The statistical test may prove to be significant but may also hide some interesting patterns in the data.

I(c) Limitations and evaluations

Of all the criteria in Unit 6 this is the weakest section for almost all students. If you are aiming for the highest marks then all your efforts in the first two parts of Interpreting and evaluating will be undermined if you have not developed evaluation skills beyond weak GCSE level.

This is a key test of your ability to be scientifically objective. At this level simple statements about problems or vague references to something that might have happened are likely to gain little credit. As in Unit 3, evaluation means looking for evidence on which to base a balanced review of the whole investigation and the validity of the conclusions. For the highest marks you must consider both the limitations of your results and what this means for the overall conclusions you have discussed in I(b).

Examiner tip

When discussing limitations, do not suggest basic practical errors. Admitting that you cannot carry out basic tasks in a competent manner will not support evidence for higher marks at A2.

If this is meant to be an objective review of evidence leading to your opinion then it is not a good idea to begin sentences with phrases such as 'I think my results are reliable...'. Present your evidence first, then summarise the reasons for your final opinion.

Limitations of results

Sample numbers

In most investigations suggestions concerning sample numbers gain little credit because the number of samples should be a prime consideration at the planning stage. Investigations at a lower level may contain statements such as 'More repeats should have been performed'. Such statements need evidence and explanation. Just collecting more data does not automatically improve the reliability of a poorly designed investigation or the validity of the conclusions made.

Variability in the data

This is a good place to start when seeking evidence for an evaluation. Begin by looking carefully at your raw data. In theory, if you have controlled repeats then each repeat should be identical. This is unlikely, so what could be causing this variability *no matter how carefully you carried out your procedure*?

Knowledge check 27

What is meant by 'range of data'?

Knowledge check 28

If there is considerable overlap between two sets of data, what might this tell you about the possibility that these are significantly different?

Outlier is another name for an anomaly — a reading that is significantly different from the rest of the data or shows a different trend from the rest of the data.

These are some of the types of evidence you might consider, for example:

- What is the range of your data? How large is it compared to the mean?
- Can you calculate the standard deviation? How large is this compared to the mean? When comparing two sets of data, is there a big difference between the standard deviations?
- Have you found any anomalies? What did you do about them? Will this affect the reliability of your data?
- Does your graph provide any evidence of an overlap between data? How might this affect the confidence you have in your conclusions?
- Does your graph provide any evidence of **outliers** or anomalies that you have not recognised previously?
- Are your data positively or negatively skewed? Could there be a reason for this and is this an effect of your sampling?
- Is there any pattern in your data that suggests that your conclusion might not be the full explanation?

Which of these is most important will depend upon your investigation. You need a different approach when looking at measurements of a single feature of organisms in two different areas than when looking at repeats of a controlled laboratory experiment.

Genuine difficulties with measurements

You will only gain full credit for considering problems with measurement of the main independent and dependent variables if you have made a reasonable effort in the planning and execution of your investigation. However, there are genuine difficulties with the techniques available to you which, no matter how hard you try, will still have major drawbacks. The problem of measuring light intensity in the field is a good example as is the problem of assessing 'growth' of plants in the laboratory. In both these cases you would be expected to link the evaluation to your data. Is there any evidence that light readings were extremely variable or that whatever was used to assess growth was unreliable?

Systematic and random errors

Both these terms are important features of the 'How Science Works' criteria and so are considered important in the assessment of Unit 6.

Random errors

As the name suggests these are errors that could occur when taking measurements and they are unpredictable. Their most important feature is that they can affect each individual measurement in a different way, so you cannot be sure how far from the true reading your measurement might be.

Random errors occur in almost any measurement and are almost impossible to eliminate entirely. In some cases this could be a serious problem. For example, measuring soil depth by pushing a spike into the ground to its maximum depth is extremely difficult to control. Problems such as standardising how much force is applied or whether the spike might encounter a large stone, make this procedure prone to random error. In other cases, such as using an electronic balance that is carefully zeroed each time, the random error is unlikely to be large enough to have a significant effect.

Systematic errors

These errors are consistent for each measurement and can be quantified. The most obvious source of such errors is poor calibration of the measuring instrument. For example, if a stopwatch is faulty and always runs slowly by 10% then each reading would be 10% lower than the true value. All the timings would be affected by the same amount, but the overall pattern and any comparisons would not change.

Detecting random and systematic errors

Random errors are likely to be a major source of variation in your data, so you should begin to think about how things were measured. Your data will give you some clues about which measurements are the most variable.

Systematic errors are much more difficult to detect as they might not be obvious in your data. However, it is worth considering at the planning stage where systematic errors might occur, so that they can be avoided by simple checks. The stopwatch example above can be checked easily by timing something with two watches and comparing them. If temperature is critical to your investigation you can do the same with thermometers (laboratory thermometers can be surprisingly variable).

Limitations of conclusions

Correlations and causation

This is a part of the content of Unit 1 and, perhaps because it is AS material or it seems a long time ago, this idea is ignored consistently by students when evaluating conclusions. It is an important 'How Science Works' criterion and therefore should be a priority, where relevant, in your report.

Examiner tip

You cannot gain credit for simply stating 'correlation does not mean causation' in your report. You must explain how this problem is linked to your hypothesis and data.

Knowledge check 29

Lyme disease is transmitted to humans by tick bites. Ticks are often found in bracken and other plants on hillsides throughout the country. Name two possible human activities that could give false correlations with Lyme disease.

The principle is simple. Just because one variable changes when a second variable changes does *not* mean that this variable causes the change. The reason for this is that many variables are linked closely with each other and so you could be observing a secondary link, not the primary cause.

Here is a simple example. It is known that serious liver disease can be caused by excessive alcohol intake. However, people who drink to excess are also more likely to show other trends. Hence, it is quite possible to show that there is a positive correlation between the amount of peanuts or crisps consumed and liver disease. This gives the totally wrong impression that eating peanuts or crisps causes liver disease. Therefore you are expected to show that you understand that demonstrating a correlation does not necessarily show that one thing causes another and your discussion should show that you understand what other links might be possible. This is another example of the need to use cautious language when considering conclusions.

Comparing your data with similar work

It might be possible to research other work that shows a similar pattern to your own. This could be used to suggest that you have greater confidence in your own conclusions, but this needs treating with great care. If you are looking at published work it is unlikely that the methods used will be similar to your own, so may have limited relevance.

Quoting similar work of your friends or students from previous years might well indicate that you have simply carried out an identical investigation. This is likely to limit your marks, so is best avoided.

Communicating

What is expected in this criterion?

- Your report is presented clearly and logically in a scientific way. C(a)
- Your data are presented using well-chosen graphs, tables or diagrams. C(b)
- Spelling, punctuation and grammar are correct and sources used are identified clearly in a well-constructed bibliography. C(c)
- You have used at least one clearly identified scientific journal and evaluated several of the sources you have used. C(d)

This criterion has a wide range of requirements so you must check carefully that you have covered everything that is needed.

C(a) Presentation of your report

The first requirement here is that your report is divided into sections that follow a logical sequence. You should avoid repetition and over-elaboration, which will make your report much too long.

Including an abstract

Most scientific papers begin with an abstract so it is a good idea to include one at the beginning of your report.

Abstracts are an important way in which scientists find scientific papers that might be relevant to their work. Hundreds of papers might be published each month, so to make the task easier each has an abstract containing the following:

- a brief summary of what is being investigated
- what methods were employed
- what the main findings were

Abstracts are usually less than 200 words.

This means that it is possible for scientists to pick out papers that are relevant to their work without having to read each one in detail, only to find it was not useful.

You may have come across abstracts in your own research as many are available free of charge and provide useful information without too much complicated detail.

An example of an abstract from a core practical

This investigation was to compare the tensile strength of plant fibres from celery and flax. 10 cm samples of sclerenchyma tissue of standard thickness as estimated under low power magnification were extracted from the mid-portion of stems of celery and flax. The samples were clamped securely at the exact length and 0.25 g weights were added carefully until the fibre broke. Flax fibres were found to have a significantly higher tensile strength as shown by a Mann–Whitney U test at the 5% confidence level. It is suggested that this difference is caused by the different structure of the sclerenchyma tissue in both plants.

Points to consider in presentation

- Use subheadings that match the criteria where it is logical to do so.
- Only include illustrations that have a clear purpose.
- It is not necessary to repeat all the details of your method in both the trial phase and the final account. Where full details have been given just describe the changes made as a result of the trial.
- If you have large amounts of raw data place these in an appendix and use summary tables in the report.
- Your report should contain only a small number of graphs.
- Think carefully in planning to avoid the need for multiple statistical tests.
- Make sure you have a clear system of referencing your sources in the text to show exactly where they have been used and that this is clearly linked to your bibliography.

C(b) Presenting data

Graphs

Selecting the correct format for a graph and presenting graphs accurately is often done badly, so this section could improve your marks significantly.

First, consider why you are including a graph. You will lose marks if you do not, but there are more important scientific reasons. A graph is meant to be a pictorial representation to help you analyse the main trends and patterns in your data, so it must be accurate and clear.

Some basic rules for drawing graphs

- Make sure that you include the main summary graph that links directly to your hypothesis.
- The independent variable should normally be on the horizontal axis and the dependent variable on the vertical axis.
- Each axis should be clearly labelled, including units, and have a correct scale.
- You may use any method you choose to draw your graphs. What will be assessed is the final product. There is no merit in drawing a graph using a computer program such as Excel if you are unable to do this accurately.
- All plotted points must be shown clearly.
- All graphs should be of reasonable size with axes of suitable length matched to the range of data. Avoid reducing the vertical axis of graph just to fit it neatly on a page when this distorts the pattern the graph displays.

Examiner tip

There should be a maximum of one or two graphs in your report. Multiple graphs are normally unnecessary. Choose carefully rather than adding more.

Main graphical formats

Line graph

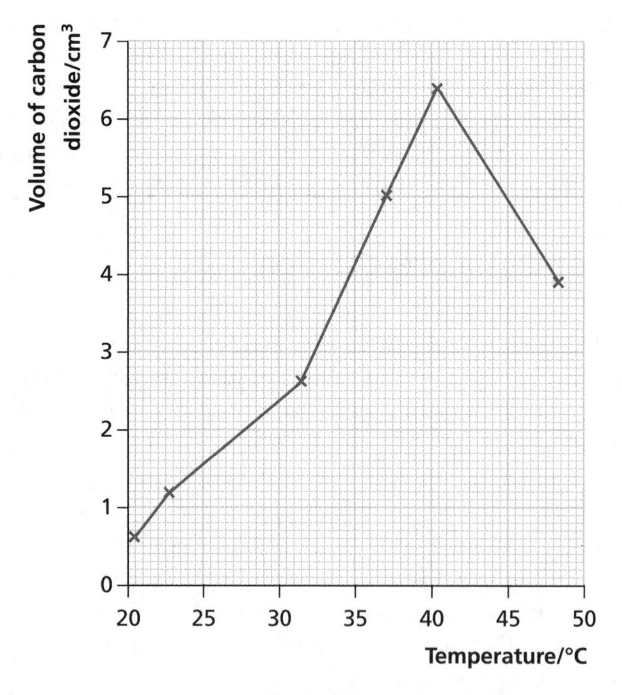

Figure 10

Both axes in Figure 10 are continuous variables. This means that it is possible to have any measurement on the temperature scale between 20°C and 50°C, or any volume measurement.

It is advisable to join points with a straight line as attempting to draw acceptable curves freehand is difficult and you do not know exactly what happens between each point.

Bar chart

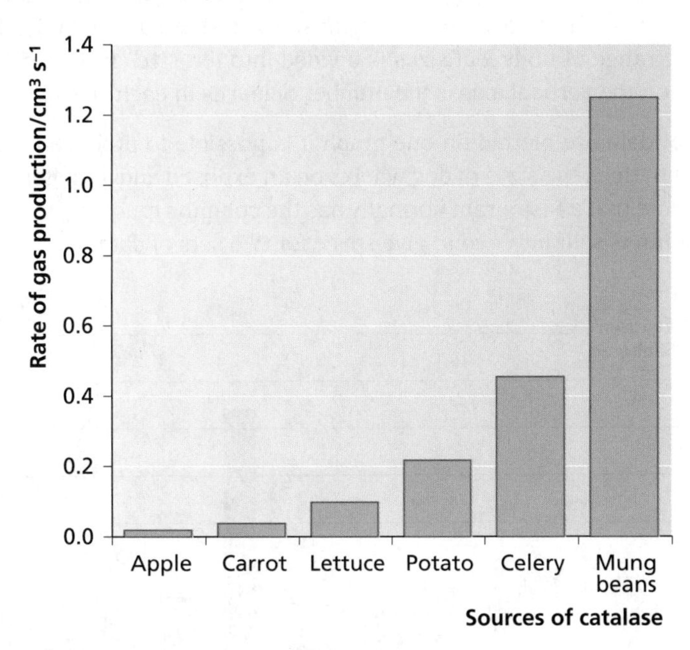

Figure 11

Bar charts (Figure 11) are used when the independent variable on the horizontal axis is not a continuous scale but a distinct category. To show this, the columns do not touch each other. A bar chart can also be the simplest way to display two means.

Histogram

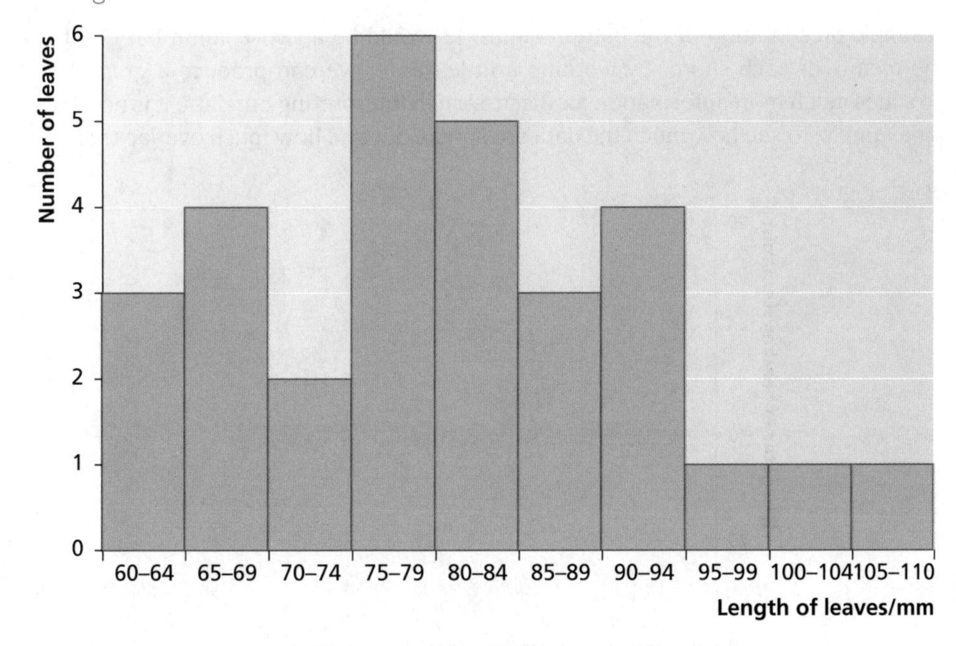

Figure 12

A histogram is also drawn with columns but the horizontal axis is often the data from the dependent variable measurements organised into size classes. In the example in Figure 12 the range of holly leaf sizes is divided into ten size classes of 5 mm each. The number on the vertical axis is the number of leaves in each size category.

If two sets of data are plotted on one graph it is possible to produce a comparison (see Figure 13). Here, the sizes of dog whelks on an exposed and a sheltered shore are compared. Note that a histogram normally has the columns touching, but in this case a single column is split into two to give space for two sets of data.

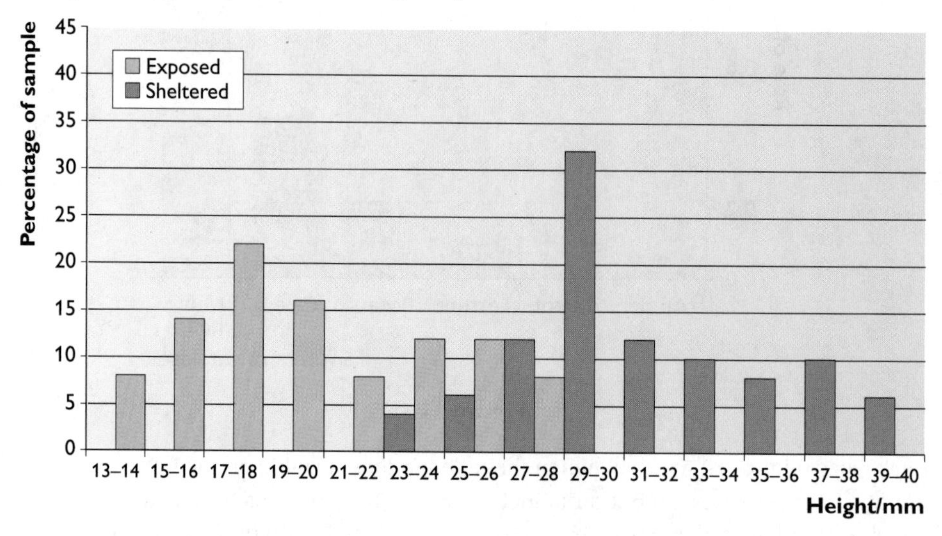

Figure 13 Histogram to show variation in shell height in dog whelks from sheltered and exposed shores

Examiner tip

Do not use 'sample number' as an axis because this may be scientifically meaningless. Random samples from two different sites cannot be paired together just because you have numbered them the same.

A simple presentation of the data in Figure 13 would be a two-column bar chart of the means of each shore. By working a little harder we can produce a graph that provides much more information for discussion in interpreting our data. It is possible, for example, to see how much the data are spread out and how much overlap there is.

Scatter graph

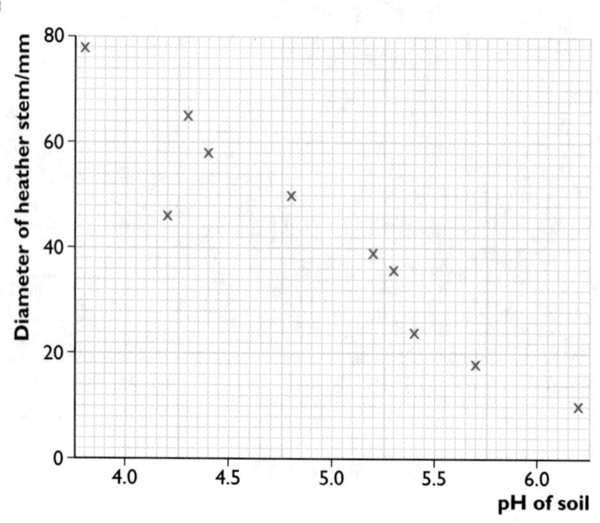

Figure 14

Scattergraphs (sometimes called scattergrams) are used when investigating a correlation between two variables. The data are plotted as individual points. This can often highlight possible anomalies for further investigation, such as the low diameter at pH 4.2 shown in Figure 14.

The pattern of points also indicates the general trend of any correlation present as shown in Figure 15.

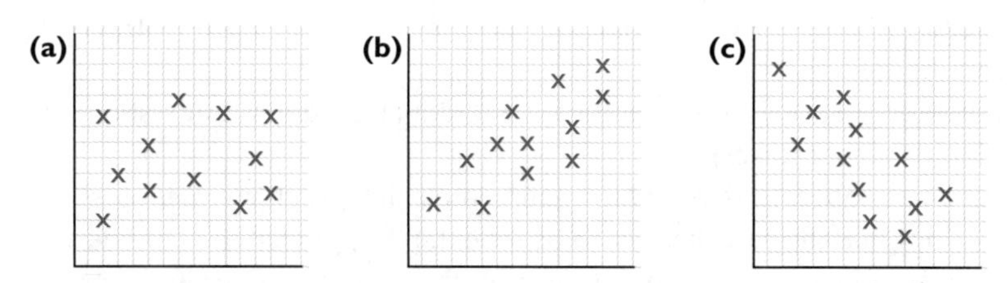

Figure 15 Three possible trends on a scattergram: (a) no correlation; (b) positive correlation; and (c) negative correlation

Examiner tip

Don't attempt to draw a 'guesstimate' straight line on a scattergraph. It is better not to include a line and to rely on a statistical test for evidence of a significant correlation than to introduce a biased attempt.

'Box-and-whisker' plot

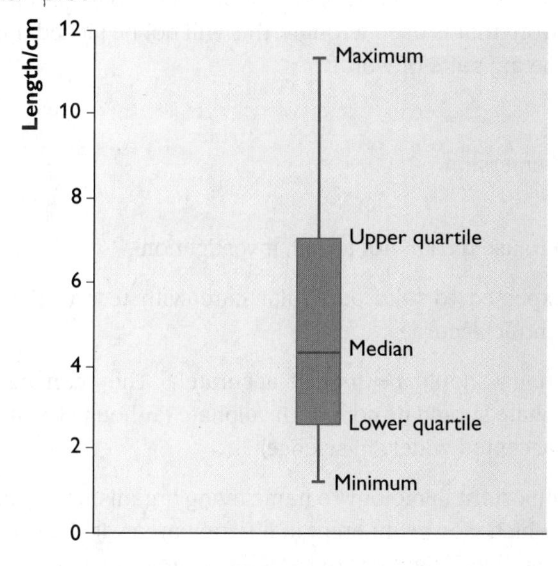

Figure 16

Box-and-whisker plots (see Figure 16) are most often used to represent skewed data but they also provide lots of information about the data. Plotting two box-and-whiskers side by side can also be useful in comparing data sets.

To find the values to plot, all the measurements are ranked in a list. The list is then divided into quartiles. The lower quartile boundary is the measurement that has 25% of the sample. If there were 16 readings then this would mean moving up the rank order until you have the fourth measurement. The upper quartile boundary is when

Knowledge check 30

Look at the 'box-and-whisker' plot in Figure 16. Suggest one reason why this shows the data are not a normal distribution. Are the data negatively skewed or positively skewed?

you have 75% of the sample (the 12th measurement in our example). The 'box' is now drawn as a column using the upper and lower quartiles on the vertical scale. The 'whiskers' are added as lines with a small cross-bar to indicate the highest and lowest measurements and the median is drawn as a line in the box.

Tables of data

As we have seen in the section on observing, you should use SI units wherever possible.

- Tables need to be designed so that they display the data clearly with consistent significant figures.
- Column headings should describe the data and include units (units should not be included in the other boxes).
- If data are not too extensive then manipulated data such as means should be included alongside the raw data (such as repeats) from which they are derived.
- The main table must give a clear summary of the data used to draw all graphs.

C(c) Spelling, punctuation and bibliography

Always use a spellchecker as you write your report, but do remember that where you type in a correct word that is used wrongly, this will not be picked up. Take particular care with the following pairs of words:

- where and were
- effect and affect
- immersion and emersion
- there and their

Knowledge check 31

Write one sentence containing the words effect and affect to illustrate the difference between them.

You are expected to use a trial, not a trail, investigation!

At A2 you are expected to take particular care with technical terms. These are important for scientific accuracy.

Chemical compounds should be named accurately. They can be checked easily. Hence copper sulphate should be copper(II) sulphate (although the American spelling of sulfate is now accepted widely in science).

It is particularly important in biology to name living organisms correctly and to avoid common names, which may be different in different areas. It gives a particularly poor impression to investigate a plant or animal whose name you cannot spell correctly or give in the correct format. Biological names are usually given as follows:

- Example: common dandelion
- Latin name: *Taraxacum officinale*
  - written in italic script
  - genus name starts with a capital letter
  - species name starts with a small letter
- Abbreviation: *T. officinale*

Knowledge check 32

What are the correct biological names and abbreviations for (a) broad bean (b) house sparrow (c) brown trout.

These names are like our own, they do not need a definite article 'the' in front of them. Just as you would not write the John Smith, neither should you write the *T. officinale*.

C(d) Evaluating sources and using scientific journals

Scientific journals

Basic information on scientific journals is provided in the introduction to this book under 'How Science Works'.

If you have undertaken research for Unit 3 you should be familiar with some of the main scientific journals. Many are easy to recognise by their names as they start with 'Journal of...'. So you will meet, for example, *The Journal of Molecular Biology* or the *Journal of Infectious Diseases*. Others, such as the most prestigious American journal *Proceedings of the National Academy of Science* have historically-based titles that are not so obvious. Some are slightly different because they accept brief reports before publication of the fully detailed peer-reviewed paper. You may come across *Nature* as one of these.

Others, such as *Biological Sciences Review*, that are not journals are acceptable as they have strong academic content.

It is worth repeating here that you must name the journal in the correct format in your bibliography not just copy the internet address where you may have found it.

Evaluating sources

You will have used this skill in Unit 3 and the information and exercise given in section 3.3 (pp. 21–22) need to be revised carefully as this section is often weak and the limiting factor in this criterion. Remember there is little credit for repeating basic phrases such as 'I cross-checked this information with other sources' without evidence of exactly what information you are referring to and where exactly you found confirmation.

Knowledge check 33

What is the correct format for a journal reference in your bibliography?

Sample Exercises

Exercise 1 Choosing a title

A group of students selected the following titles for their Unit 3 reports:

1. **Stem cells**
2. **Culling badgers is the only way to prevent the spread of bovine TB**
3. **Developing a vaccine to prevent malaria**
4. **Reducing carbon dioxide emissions by increasing the use of wind farms**
5. **Producing strains of bacteria to digest marine oil spills**

(a) For each title, state whether you feel it is suitable or unsuitable. Explain your decision.

(b) For each unsuitable title explain how it could be improved to meet the criteria.

(a) **1** Unsuitable — much too vague and does not identify a problem
2 Suitable — clear problem with a suggested answer; biological background and data are available; several alternatives to discuss
3 Suitable — malaria is a serious problem; lots of work going on to develop a vaccine (not yet successful) with lots of alternatives
4 Unsuitable — solution is not biological: it is largely political and economic
5 Suitable — some good biology in breeding strains of bacteria often with genetic manipulation; clear problem with alternative solutions

(b) **1** Could become suitable with a focus on an example of research into treatment of one condition using stem cells
4 Probably best avoided: many solutions are political rather than biological and could lead to long descriptions of greenhouse gases rather than the biological research taking place

e Take care if you meet questions like these in a unit test. Each has two distinct sections, so your answer should be divided into two distinct parts.

Exercise 2 Researching

Imagine you are investigating the effect of heating on vitamin C in foods.

Type 'vitamin C' into your normal search engine and make a list of the websites on the first page of the search.

Now repeat this exercise but using double quotes type in "effect of heating on vitamin C".

(a) What are the main differences in your lists?

(b) What does this tell you about the importance of selecting search criteria carefully?

(a) The first sites for the vitamin C search are highlighted and are adverts for the sale of vitamin C products. These are sponsored links paid for by the advertisers. The rest is a mixture of sites giving information on vitamin C and irrelevant references on YouTube.

(b) Single word searches often yield a large number of irrelevant sites. This wastes time and makes it more difficult to find relevant information. Make sure your searches are closely related to your report title.

e Whenever you look up information on a website, always keep asking yourself why that website is available, who is paying for it and where does the information come from.

Exercise 3 Summarising methods

The aim of this question is to practise summarising biological methods in a concise way. Your answer should aim to be between 150 and 200 words. You do not need to include the results or conclusions.

For each of the following explain *one* method by which biologists have found answers or solutions to the following problems.

(a) Does DNA replicate by a conservative or semi-conservative method? (8 marks)

(b) Should living things be divided into just three major domains? (6 marks)

Key points are italicised in the answers. Check your answer by awarding 1 mark for each point. A total of 6 marks would be a good answer for (a) and 4/5 marks a good answer for (b).

(a) When DNA is placed in a *centrifuge tube* with a strong salt solution and *spun very fast* then *a band forms* in the tube that can be detected. If the DNA is made heavier by *labelling it with ^{15}N* then a different band forms lower down the tube. Dividing *bacteria were grown in a ^{15}N medium* so all their DNA was 'heavy'. They were then *transferred to a normal 'light' ^{14}N medium* and sampled at short time intervals to see what happened to the DNA when tested in the centrifuge. The appearance of a *new band of DNA* that was *halfway between 'heavy' and 'light'* demonstrated that the DNA had one new 'light' strand and one 'heavy' strand, so it was replicating semi-conservatively.

(b) Traditional classification uses mainly external features and some simple biochemistry. In 1977, Carl Woese proposed a new kingdom of ancient bacteria that he called *Archaea*. This was based on an *analysis of RNA sequences*. This involves using chemical techniques to find the exact base sequence of the RNA. At the time this was a very slow process but he found a group of *base sequences that were missing from this group compared with all other bacteria*. His idea was not generally accepted until the sequencing of DNA and RNA became much more automated. In 1996 the *full gene sequence of one of the ancient bacteria* was determined and this proved conclusively that the Archaea were very different from other bacteria and *a separate domain*.

These are two examples taken from AS specification material. It is easy to find much more information about each, some of which might be relevant. In general, however, if you are accurate, the answers should not need to be extended into several pages, as there is more important evidence of your skills to be provided later.

Exercise 4 Evidence from data

e A student chose the problem of conserving rare Amazon river turtles as the title of their Unit 3 report. He described several conservation measures undertaken by biologists in Brazil.

Turtles lay their eggs in holes dug in sand on the riverbank. The eggs are then buried to protect them from predators. The team of conservationists established protected sites on the riverbank to prevent disturbance by human activities and predators and counted the numbers of nests in each site. The student found the following evidence of the success of their work.

Table to show total numbers of protected turtle nests

Year	2000	2005	2006	2008	2009	2010	2011
Giant South American river turtle	18	12	16	9	21	55	133
Yellow-spotted river turtle	81	124	223	191	32	181	379

(a) What exactly do these data show? (1 mark)

(b) How reliable are these data? (3 marks)

(c) The aim of the conservationists' work was to increase the population of these rare species. Do these data give valid evidence that their work is achieving this aim? (4 marks)

(a) The total number of protected nests in the conservation sites only. ✓

(b) The data are only partially reliable. We would need to know a lot more about counting and how often the sites were observed etc. ✓
The numbers of each show a lot of variation especially the large falls in 2008 (giant turtle) and 2009 (yellow-spotted turtle). ✓
The data might be more reliable from 2009–2011 — there is a consistent upward trend that has lasted for 3 years. ✓

(c) These data show only the total number of nests; the total population numbers are not recorded. ✓
The population may remain low if some other factors are causing high mortality of newly-hatched turtles or of adults in the river. ✓
The data show only the number of nests within the protected areas. There may be a decline in other non-protected areas or turtles may be migrating to the protected areas. ✓
However, the increase in the total numbers of nesting turtles does give some valid indication that the whole breeding population might be increasing and that protecting sites is a successful strategy. ✓

e The purpose of this exercise is to illustrate what is meant by considering validity and reliability and how using evidence is essential to forming judgements. In this case, thinking carefully about exactly what the data do and do not show and the danger of jumping to conclusions without careful analysis. However this does not mean that these data are irrelevant. If discussed carefully, they could gain full credit in both 1.2 and 1.3.

Note that parts (b) and (c) could be used as evidence for criterion 3.3 as they begin to evaluate the evidence quoted.

Exercise 5 Benefits and risks

Evaluating

(a) Look back at your work in Unit 1 on cystic fibrosis. One possible solution to this problem is gene therapy.
- **(i) List two benefits of this solution.**
- **(ii) List two risks or drawbacks associated with this solution.**
- **(iii) Look carefully at the two lists and explain in your own words whether you feel the benefits outweigh the risks and why.**

(b) Table 1 gives some data on the work of zoos with cheetahs.
The problem being addressed in this report is 'How can the decline in numbers of wild cheetahs be prevented?'
- **(i) What are the obvious benefits of the zoos' captive breeding programmes?**
- **(ii) Explain why this solution does not necessarily provide an answer to the main problem.**
- **(iii) It has been suggested that the captive breeding programme could lead to the reintroduction of cheetahs to their wild habitats. Why is reintroduction so difficult?**

Table 1 Numbers of cheetahs in captivity

Year	1999	2000	2001	2002	2003	2004	2005	2006
Numbers of cheetahs	1290	1315	1371	1340	1349	1382	1433	1408

(a) **(i)** It has the potential to provide a long-term cure rather than repeatedly treating symptoms. This would give sufferers a large increase in quality of life.
A single treatment would be more cost effective than years of expensive therapy.

(ii) Transferring genes into cells could have unforeseen effects. No trials show what the long-term effect might be.
Gene therapy only treats existing cells, when these die off they are replaced naturally with cells containing the faulty allele. Hence the effect may be short-lived.

(iii) Gene therapy for cystic fibrosis has enormous benefits and could be an excellent solution. At the present time, unfortunately, it is not a true solution. This is because the effects it produces only last for a short time, which means that the patient has to go back to conventional treatments or undergo repeated gene replacement. Depending upon how the gene is being delivered using different vectors, such as viruses, this might increase risks. However, stopping such research would mean that there would be no hope for the future. Therefore, such therapy is likely to continue in the hope of refining methods and finding ways of getting correct copies of the allele into the cells that divide to replace the epithelial lining of the lung. This would provide an ideal permanent solution.

e Note that in part (iii) there is an attempt to use information to come to a balanced conclusion that could be regarded as evaluation. This can be achieved in short paragraphs, not long descriptions.

(b) **(i)** The number of cheetahs has increased.
(ii) This number is the number of cheetahs in captivity. It does not increase the number of cheetahs in their natural habitat.
(iii) Cheetahs live on open plains and capture their prey by careful stalking followed by a short burst of rapid acceleration. These skills are learned from adults; animals bred in captivity will not have acquired either these skills or other survival strategies.
Simply introducing more cheetahs into the wild does not address the questions of why they are declining and what can be done to change the underlying causes of their decline.

e In this case the value of the solution is more debatable and its drawbacks could lead to possible alternative solutions for criterion 2.3. However, the principle has not changed. There is critical comment on the effectiveness of the solution using reasoned argument, which meets the requirement to evaluate.

Exercise 6 Evaluation of sources

The following is an evaluation of a source given by a student. It was awarded 1 mark. Rewrite the evaluation in a form that would be awarded the full 2 marks.

Source: Crick, F.H.C., Barnett, L., Brenner, S., and Watts-Tobin, R. J. (1962) 'General nature of the genetic code for proteins', *Nature* 192, 1227–1232

'This is a reliable source because it is a peer-reviewed journal and it is written by some well-qualified scientists. It might not be reliable because it is now very old.'

This is a journal that has been peer-reviewed. This means that it has been checked carefully by other scientists and that the research has been carried out in a valid way. They also have to agree that the conclusions are valid before the paper can be published. Two of the researchers are Nobel prize-winning scientists. Although it was written in 1962 the basic principles of genetic coding have been confirmed by further research. However, there are additional details of this process that have been discovered since then, so care is needed when using this paper. I feel that this evidence makes this a reliable scientific source but that a more recent source would be needed for more details.

e The original evaluation gives little evidence on which to base an opinion. Some statements suggest this is along the right lines but a lack of discussion means that there is some doubt about the student's understanding of the points raised. This is an important point relevant to all of the criteria. A simple mention of a relevant point might gain credit but typically there will be twice as many marks available for adding a clear explanation or discussion.

Exercise 7 Lessons from core practicals

Try this simple test:

(a) All enzymes denature when heated to 50°C or above — true or false?

(b) All enzymes have very narrow optimal pH ranges — true or false?

(a) False — many bacteria and algae live in hot springs with temperatures above 50°C
(b) False — many enzymes have a range of pH values over which they operate efficiently

e These are two examples where many textbooks have similar graphs showing the principle of optimum temperature and optimum pH. However, there are many variations some of which are important biologically. It is not uncommon for students to assert that 37°C (human body temperature) is the optimum for an enzyme such as catalase extracted from celery! This is a good example of where you need to think more carefully, not just regurgitate information.

Exercise 8 Research and rationale

A student is investigating the idea that ivy leaves in the shade have a greater surface area than those in a well-lit area. In the research-and-rationale section, she included the following pieces of information:

- **three pages describing the biochemistry of photosynthesis**
- **two pages with photographs describing ivy**
- **two pages describing the woodland where the investigation was carried out**

This section was awarded 6 marks out of 11.

(a) Why might the photosynthesis information be too long and only given limited credit?

(b) Is the information on ivy useful here?

(c) How might the woodland information be useful in other sections?

(d) What vital piece of research is missing from this list?

(a) The biochemical details of photosynthesis cannot be tested or measured in this investigation. A *summary* is relevant because it is an important way in which light affects plants.

(b) A knowledge of ivy as an epiphyte and the shape of its leaves would be useful in planning and selecting sample sites.

(c) The location of ecological investigations is important. This information might be used to consider what is meant by 'well-lit' and 'shade', how this would change throughout the day and, later, in evaluating.

(d) The missing piece of research is exactly how light could make leaves grow bigger. More food from photosynthesis is a simplistic idea. Why would this not mean just more leaves rather than bigger ones? There is plenty of information available to show how light is detected and can change the quantities of plant hormones, hence overall shape or size.

e This investigation is also an ecological exercise and some consideration of the value of this response to the ecological niche of the plant could be included. This is an example of how an average mark could become a very good mark by thinking more carefully about the hypothesis under test. Some consideration of how changes in leaf area might be brought about would move this to 9 marks or more.

Exercise 9 Trial investigations

For each of the following investigations suggest simple trials that would be useful in establishing a reliable method.

(a) Investigating the size of leaves in light and shaded areas

(b) Growth of plants is inhibited by common freshwater pollutants

(a) The main variables here are:
- independent — light and shade
- dependent — size of leaf

What is the best measure of size? Record some data to check if measuring width be a good representation of area? What would be the best way to define exactly what was measured to make it consistent each time? Does it make a difference where on the plant the leaf is sampled? How could this be controlled? How can I define light and shade or make my light measurements reliable? Is there any effect from surrounding trees?

(b) The main variables here are:
- independent — concentration or presence of pollutant
- dependent —'growth' of plant

Why might just measuring height or length be a poor method, particularly for seedlings? Is it possible to untangle seedling roots to measure them reliably? How long do the plants need to grow before a reliable set of measurements can be taken? Research the common concentrations of pollutant. Test for a suitable range of concentrations to use.

e At A2 ways of measuring 'growth' need to be researched carefully.

Exercise 10 Descriptive statistics

A student carried out an investigation to test the idea that the leaves of a holly bush growing in a sunlit area would have a smaller surface area then those growing in a shaded area. She measured the light intensity and the surface area of a random sample of 15 leaves from each area.

The table below shows some of the data she collected.

Area of leaves in sunlight/cm^2														
17.1	18.0	19.5	20.0	17.9	21.6	18.1	17.5	19.8	20.1	18.9	17.7	17.4	17.9	18.2

(a) Write a suitable null hypothesis for this investigation.

(b) Are these data normally distributed? Explain your answer.

(c) For these data, calculate:
(i) the median
(ii) the mean
(iii) the standard deviation
(iv) the mode

(a) There is no significant difference between the surface area of leaves growing in the shade and those growing in full sunlight.

(b) No. The data are negatively skewed as there are ten readings in the range 17.0–18.9 but only five readings in the range 19.0–21.9. This could be plotted as a histogram to show the difference more clearly.

(c) (i) median = 18.1
(ii) mean = 18.65
(iii) standard deviation = 1.28
(iv) mode = 17.9

Exercise 11 Applying a statistical test

A student forms the hypothesis that there is a significant correlation between the moisture content of soil and the abundance of common rushes.

He collected data from 14 random samples and calculated that the r_s value was 0.653. A table of critical values for r_s is shown below.

Number of pairs of data	Significance level			
	10%	5%	2%	1%
12	0.506	0.591	0.712	0.777
14	0.456	0.544	0.645	0.715
16	0.425	0.506	0.601	0.665

(a) What is the critical value of r_s for these data at the 5% significance level?

(b) Compare the calculated vale of r_s with the critical value and explain what conclusions can be made.

(a) 0.544

(b) The calculated r_S value is higher than the critical value so the null hypothesis is rejected. There is a significant correlation between abundance of common rush and soil water content (at the 5% significance level).

e Remember it is important to include the word 'significant' when describing your conclusion.

Exercise 12 Using information to interpret data

The following is a brief extract from an interpreting section of an investigation into the difference in volume of limpets found on an exposed shore compared with those found on a sheltered shore.

'At the 5% confidence level my t-test shows that limpets on the sheltered shore have larger volumes than those on the exposed shore. This agrees with my initial hypothesis and my research. Sources (3) and (4) both suggest that the most likely explanation is that limpets on sheltered shores are less subject to strong wave action throughout the year. Source (4) suggests that this means their muscles expend less energy contracting to prevent them from being dislodged and therefore more energy is available for growth. Sources (5)(6) have demonstrated that larger limpets are more likely to be swept away on exposed shores, which would also limit their size.'

Identify three features of this extract that would indicate that it could be a part of a high-scoring section.

1. The extract makes good use of source material that is referenced clearly.
2. The use of words such as 'suggests' and 'most likely' indicate a cautious objective approach rather than unquestioning assertion.
3. The extract shows an understanding of the fact that there may be more than one explanation and the effect might be a combination of different factors.

e You are expected to include more detail than this extract shows, particularly of the biological information given by the sources. This is just an indication of the type of approach needed.

Exercise 13 Interpreting data in detail

The table below shows the results of an investigation into the effect of lead ions on germination of mung beans. Fifty mung beans were placed in each of seven shallow dishes with controlled volumes of distilled water and solutions containing different concentrations of lead ions. The number of beans germinated after 24 hours was recorded.

Concentration of lead ions/mmol dm^{-3}	0	0.001	0.005	0.01	0.015	0.02	0.025
Number germinated in replicate 1	48	49	36	29	21	11	0
Number germinated in replicate 2	49	49	38	26	18	9	3
Mean	48.5	49.0	37.0	27.5	19.5	10.0	1.5

(a) Use a Spearman's rank correlation test to find the correlation coefficient for these data.
(b) Use a table to find the critical value for seven pairs of measurements at the 5% confidence level.
(c) Write an accurate statement to describe exactly what this means about the link between concentration of lead ions and germination.
(d) A student researched information about the effects of lead ions and found that, in addition to the obvious inhibitory effect, there were some sources that described an effect called hormesis where very low doses of toxic substances stimulate germination rather than inhibit it.
The student claimed that her data show that this effect is demonstrated by her investigation.
Evaluate the evidence for this claim.

(a) −0.964
(b) 0.786
(c) The calculated coefficient is greater than the critical value (ignoring the minus sign) at the 5% confidence level. We can reject the null hypothesis as there is a significant negative correlation (shown by the minus sign) between the concentration of lead ions and the number of mung beans that germinate.
(d) There is only weak evidence for this claim as the germination at 0.001 mmol dm^{-3} is greater than that in distilled water. However, this conclusion is based on just one seed difference. The other concentrations show that, even in the same concentration, there is variability between the repeats that is often more than one seed. Therefore, this might simply be random variation between the 50 seeds placed in the dishes. A much greater number of mung beans would have to be tested using a much narrower range of dilute solutions of lead ions before we could say that hormesis is occurring.

e Part (d) is an example of beginning to evaluate using evidence and showing suitable caution in conclusions. In most investigations there might be more data and standard deviations to discuss, and possibly sufficient data to carry out a statistical test. This is not a recommended investigation as it is very simple, but it provides an example of where it is possible to discuss the need for more data.

Unit 6 student checklist

Criterion		✓
Research and rationale	The biological background to your actual hypothesis is explained.	
	There is a clear indication, in the text, to show where researched information has been used.	
	Researched information has also been used in planning *and* in explaining the results.	
Planning	There is a clear plan of action.	
	Sufficient data will be collected, which are matched to the statistical test.	
	There is a trial investigation.	
	The results of the trial investigation are used to amend the method.	
	The most important dependent and independent variables are carefully controlled.	
Observing and recording	Data are recorded with SI units and consistent, appropriate significant figures.	
	Data are sufficient to make a reasonable conclusion.	
	Any possible anomalies are noted, or if none, then some explanation is given.	
	Where anomalies are noted, action is taken and explained.	
Interpreting and evaluating	The statistical test is clearly set out.	
	Conclusions from statistical testing are clearly explained in my own words.	
	Results are interpreted using researched information.	
	My evaluation uses evidence from my data.	
Communicating	The report uses clear sub-headings and has a short abstract.	
	The format of graphs is correct and they are carefully selected.	
	The spelling of technical terms and names of organisms has been checked.	
	All references are in a correct scientific format and there is at least one scientific journal.	
	All my evaluations of sources are based on evidence.	

Knowledge check answers

1 Large drug companies had spent tens of millions of pounds on developing sophisticated acid-reducing drugs. Many doctors had dismissed the idea that bacteria were involved.

2 No — reducing the production of stomach acid is also important. The standard treatment is to use antibiotics along with acid-reducing therapy.

3 As soon as you add the enzyme things change quickly. The substrate is being used up and the rate changes all the time. The initial rate is the closest we can get to the correct rate with all the starting variables controlled.

4 Coffee and 'Red Bull' contain other active compounds, so it is not possible to tell whether it is the caffeine that causes the effect.

5 There are problems in using sweet, sticky honey in wound dressings. Eating honey means that many of its active compounds are changed during digestion and, therefore, will not enter the blood stream.

6 The site is full of adverts. You may also get 'pop-up' boxes with adverts. There is a disclaimer in red at the top, probably for legal reasons, and advice to talk to your doctor. (Why do doctors not prescribe garlic?)

7 **(a)** That cows produce antibodies against the virus and these are still present after six weeks.

(b) It does not show that the antibodies stopped the cows contracting foot-and-mouth disease. It does not show the antibodies would remain in the cows' bloodstreams for longer than six weeks.

8 The size of fruits can be determined by other factors such as the number of ovules fertilised and hence the number of seeds inside. Fruit development is controlled by plant hormones. The number of fruits on one bush can also have an effect.

9 **Possible reasons to support:**

- Everyone stands to benefit from advances in science, so taxes should be used to pay for it.
- A lot of fundamental research appears initially to have no direct use, but it often leads to important advances later.
- Scientific research should not be controlled by large organisations.

Possible reasons to oppose:

- There are other better ways of spending public money.
- If the research is useful then industry will support it.

10 **In favour of GM:**

- Targets pests effectively and reduces use of harmful chemicals
- Increases yields of crops and allows crops to be grown in more areas

Against GM:

- Too risky, we do not know how it might affect the environment in the long term
- Allows big companies to control advantageous developments and make excessive profits

11 **Advantages:**

- Reduces presence of *Salmonella* and so reduces risk of infection
- Makes chickens healthier and grow faster

Disadvantages:

- Increases presence of antibiotics in chicken meat
- Likely to promote development of antibiotic-resistant strains of *Salmonella*

12 Better hygiene when rearing chickens
Reduce numbers in battery cages
Clean chickens more thoroughly before sale

13 Author, Initials., Year. Title of book. Edition (if not the first edition). Place (town or city): Publisher

14 Lees, E., 2013. Edexcel Biology AS/A2 Student Unit Guide; Units 3 & 6; Practical Biology and Research and Investigative Skills. Oxford: Philip Allan

15 **(a)** Independent variable — the variable you control in your investigation. For example, it might be changing the temperature or concentration, or selecting two habitats with different light intensities.

(b) Dependent variable — the variable you measure to find out the effect of changing the independent variable. For example, it could be a rate of reaction, a change in heart rate or a difference in plant distribution.

16 Species richness is a count of how many different species are present in the sample. Species diversity, measured by the Simpson's Diversity Index, takes into account factors such as how many members of each species are present. It is a much more useful measurement in ecology.

17 Moles per decimetre cubed ($mol\,dm^{-3}$)

18 A 1% solution has 1 g of the substance in 100 cm^3. A molecule of sucrose has almost double the mass of a molecule of glucose. Therefore there will be twice as many molecules of glucose in a 1% solution. A 1 $mol\,dm^{-3}$ solution of glucose has the same number of particles as a 1 $mol\,dm^{-3}$ solution of sucrose.

19 The graph shows the mean is beginning to settle down to a consistent level by 19 samples, but we would need more than this to be certain we had sufficient to produce a reliable mean.

20 A transect is the best choice where there is a gradual change across the sampling area — for example, moving away from a pond or lake to give a gradient of soil water content or across a sand dune succession.

21 The point frame gives single readings of one individual. An open quadrat gives an estimate of a fixed area, either species counts or abundance such as percentage cover. A point frame is more likely to miss less common species.

22 A reading of 2.0 mm means that the measurement was exactly 2 to an accuracy of 0.1 mm. A reading of 2 mm suggests this accuracy is only 1 mm. (between 1.5 and 2.5 mm)

23 1 in 6 or $p = 0.167$

24 Exactly the same: 1 in 6 or $p = 0.167$. If the event is random then there is no 'memory' of what has happened before. So trying to pick National Lottery numbers on how many times they have appeared previously is pointless!

25 There is no significant difference between reaction times after ingesting caffeine and reaction times without caffeine.

26 H_0 There is no significant difference between the volumes of limpets on a sheltered shore and those on an exposed shore.
H_1 There is a significant difference between the volume of limpets on a sheltered shore and those on an exposed shore.

27 The range of data is the difference between the highest and lowest observations.

28 A large overlap begins to suggest that the difference between two populations might not be significant. Even if the statistical test indicates that this difference is significant it still means that a number of individuals from each population have the same measurements.

29 There could be a correlation between infection with Lyme disease and hiking or hill walking or even frequency of wearing shorts or boots.

30 The box is not located centrally between the two whiskers. It is towards the bottom end. The median line is not in the middle of the box.
Since the box is low down on the scale the data are negatively skewed.

31 The effect of caffeine on reaction time did not seem to affect all of the participants in the same way. (In this sentence effect is used as a noun to name the result or consequence and affect is used as a verb to influence something.)

32 **(a)** *Vicia faba*
(b) *Passer domesticus*
(c) *Salmo trutta*

33 Author, Initials, Year. Title of article. Full Title of Journal, Volume number (Issue/Part number), Page numbers.